2012

ISBN-13: 978-1477653982

ISBN-10: 1477653988

## Part I

## Part I

## Table of Contents                                         Page

0. Preface ......................................................................................... 3

I. Abstract ........................................................................................ 3

II. Introduction ................................................................................ 3

III. Differential Equation Population Models ................................... 4

IV. Multi-Age Mixture Continuum Model ....................................... 10

V. Discussion and Conclusion ......................................................... 13

VI. References ................................................................................ 13

Appendix A: Halley's Life Table ...................................................... 16

Appendix B: Survival Probability Table ......................................... 16

Appendix C: Perspectives on the Transformation or Decline of a Civilization ......... 18

Appendix D: Continuum Area Average Population Model .......... 21

Appendix E: Statistical Entropy Interpretations ........................... 22

Appendix F: Lattice Models ............................................................ 25

Appendix G: Hurst Exponent Time Series Analysis ...................... 27

Appendix H: Fractional Calculus Applied to Positive Continuous Time Linear Systems ... 28

Appendix I: Continuum Transport Balance Equation .................. 30

Appendix J: Filtering Data Approach for a Power Function ....... 32

Appendix K: Matrix Population Model ......................................... 34

Appendix L: A Finite Element Approach for a Fractional Advection Dispersion

             Model in One Dimension ............................................ 35

Appendix M: Point in Polygon ....................................................... 36

Appendix N: Multiple Purpose Planning and Analysis with Population ............. 37

# Part II

System Reliability of Nuclear Power Generation: Site Risk and

Engineering Design

Multiple Purpose Planning Analysis

On Nuclear Safety Risk of the Hope Creek Site

On Illinois Nuclear Siting Perspectives

Pre-Calculus Financial Math With Energy Applications      38

## 0. Preface

A disparate collection of mathematical ideas and methods for quantifying populations is presented in the literature. This paper attempts to achieve the multiple purposes of review, summary, and guidance to new areas to better characterize the system and offer illumination on the alternatives.

## 1. Abstract

Mathematical approaches are given for modeling the total population growth and decline, for two competing species, and for an age-dependent gender mixture model, along with the differential Lancaster equation system for conflict with a solution. A simple fractional calculus differential equation model generalizing the integer calculus models is defined and solved. Then a continuum description of both gender and age-dependent birth processes which has not been previously specified is developed. Discrete individuals suspended in an evolving continuous fluid motion is applied. Probability theory combining mortality contributions correctly for the assumption of independence is used. Dynamics of continua descriptions are provided with a concise mathematical framework to describe a simple area dependent population model. Old world civilization data are charted and a Weibull probability distribution fit for the decline of the Roman Civilization. These are complemented with Halley's first tabulation and an illustration on survival probability approaches. Finally, the lattice model for describing a dynamic system which may be analyzed by a Poincare map, Kalman filtering by fractional calculus, matrix models, and finite element fractional calculus dispersion are documented.

## II. Introduction

The mixing of ideas on *Human chemicals*, on an equivalence between the heat output from humans and their food intake understood by Antoinne Lavoisier which lead to studies of metabolism, on early application of the laws of thermodynamics by Gibbs (1839-1903) developed from James Watt's practical steam engine and on Sadi Carnot's abstraction of them to chemistry has started our enlightened understandings of human beings and civilization dynamics through new lenses . Further developments in mathematical foundations of continuum physics and chemistry has motivated this presentation emphasizing balances of population. Further formulations , on movements, and on energy changes are mentioned and to be addressed in another manuscript.

A continuum mixture of discrete individuals or entities superposed with variations in space and time of the concentrations as a number count in each subarea and time are representative concepts for models applied to both demography and chemical physics. For the mixture there are interactions among individuals, races, or civilizations including kinetic reactions of formation or growth, or diminution, modification, or destruction, as well as movements or flows, gradual diffusion as random spreading, in addition to larger scale jumps or leaps to more distant locations. People movements are accompanied by communications, information, and attributes that are transmitted faster with signaling especially by present-day technologies. Even with an error-correcting codes idea for word accuracy hybrid forms and constructs are slurred and blurred with some provocative and useful while others may be perceived as worthless for sustenance and for survival.

<u>Categories or Descriptors Classifying Movements of People</u>

Internal Migrant

International Migrant

Immigrant

Transnational Immigrant

Diaspora

Refugee

Step-migration [staged]

Migratory chain

Circular migration

[Reference: Lewellan]

## III. Differential Equation Population Models

The following sections present a selected progression of thoughts and models that have been presented in the literature on population dynamics differential equation modelling. The earliest population table approach of Edmond Halley for estimating population change in a steady-state city system is summarized in an appendix.

IIIA. Exponential Models

$$\dot{P} = \frac{dP}{dt} = \lambda \cdot P$$ , initial condition $P(t=0) = P_0$, and constant rate $\lambda = constant$,

gives for time t, $\quad P(t) = P_0 \cdot e^{\lambda \cdot t}$

In the case of a time-dependent rate function $\lambda = \lambda(t)$, integration and exponentiation obtain,

$$\int_{P_0}^{P(\tau)} \frac{dP(t)}{P} = \int_0^\tau \lambda(t) \cdot dt, \qquad ln\left[\frac{P(\tau)}{P_0}\right] = \int_0^\tau \lambda(t) \cdot dt,$$

which obtains,

$$P(\tau) = P_0 \cdot e^\Lambda, \quad \text{where} \quad \Lambda = \int_0^\tau \lambda(t) \cdot dt$$

For a power law rate function, $\quad \Lambda = \int_{\tau_0}^\tau \beta \cdot t^\alpha \cdot dt = \left(\frac{\beta}{\alpha+1}\right)[\tau^{\alpha+1} - \tau_0^{\alpha+1}]$

$$P(\tau) = P_0 \cdot e^{\left(\frac{\beta}{\alpha+1}\right)[\tau^{\alpha+1} - \tau_0^{\alpha+1}]}$$

A function referred to as an extreme value probability distribution with parameters $\alpha, \beta,$ and $\tau_0$ initially introduced by Walodi Weibull for crack lengths in metals.

IIIB. Consider the power function for the right hand side, $\quad \dot{P} = \frac{dP}{dt} = \lambda \cdot P^{m+1}$, with initial condition $P(t = 0) = P_0$, and rate function $\lambda = \lambda(t)$, Integration yields

$$\int_{P_0}^{P(\tau)} \frac{dP(t)}{P^{m+1}} = \int_0^\tau \lambda(t) \cdot dt,$$

$$\left[\frac{1}{-(m+2)} P^{-m-2}\right]_{P_0}^{P(\tau)} = \frac{1}{-(m+2)}\left[P^{-m-2}(\tau) - P_0^{-m-2}\right] =$$

$\int_0^\tau \lambda(t) \cdot dt, \quad$ One may again define $\quad \Lambda = \int_0^\tau \lambda(t) \cdot dt$ for more concise notation.

$$P^{-(m+2)}(\tau) = -P_0^{-(m+2)} - (m+2)\int_0^\tau \lambda(t) \cdot dt,$$

$$P(\tau) = \left[-P_0^{-(m+2)} - (m+2)\Lambda\right]^{-\frac{1}{m+2}}$$

for duration from time zero to tau.

In the case of the power time-dependent rate function $\lambda = \beta \cdot t^\alpha$,

$$\Lambda = \frac{\beta}{\alpha+1} \cdot \tau^{\alpha+1} \quad \text{and} \quad P(\tau) = \left[-P_0^{-(m+2)} - (m+2)\frac{\beta}{\alpha+1} \cdot \tau^{\alpha+1}\right]^{-\frac{1}{m+2}}$$

For m=-1
$$P(\tau) = \left[-P_0^{-3} - \frac{3\beta}{\alpha+1} \cdot \tau^{\alpha+1}\right]^{-\frac{1}{3}}$$

## IIIC. Logistic Growth

Here, the population change is assumed to be quantitatively described by the differential equation,

$$\dot{P} = \frac{dP}{dt} = (\lambda_0 - \gamma_0 P) \cdot P \quad \text{or written} \quad \frac{1}{P}\frac{dP}{dt} = \lambda_0 - \gamma_0 P, \quad \text{with parameters} \quad \lambda_0, \gamma_0 > 0, \text{ where}$$

$P(t)$ and $P(0) = P_0$  Separating variables and integrating both sides by a partial fraction expansion,

$$\int_{P_0}^{P(\tau)} \frac{dP}{(\lambda_0 - \gamma_0 P) \cdot P} = \int_0^\tau dt \quad, \quad P^0 \equiv \frac{\lambda_0}{\gamma_0}$$

$$= \frac{1}{\gamma_0}\int_{P_0}^{P(\tau)} \frac{dP}{(P^0 - P) \cdot P} = \frac{1}{\gamma_0}\left[\int_{P_0}^{P(\tau)} \frac{dP}{P^0 P} + \int_{P_0}^{P(\tau)} \frac{dP}{P^0(P^0 - P)}\right]$$

$$= \frac{1}{\gamma_0 P^0}\left[\ln\left|\frac{P(\tau)}{P_0}\right| - \ln\left|\frac{P^0 - P(\tau)}{(P^0 - P_0)}\right|\right], \quad \text{or altogether}$$

$$\gamma_0 P^0 \tau = \ln\left|\frac{P(\tau)(P^0 - P_0)}{P_0(P^0 - P(\tau))}\right|$$

Therefore the population from changes over the time interval accumulates to,

$$\mathbf{P}(\tau) = \frac{P_0 P^0 e^{\gamma_0 P^0 \tau}}{\left[P^0 - P_0\left(1 - e^{\gamma_0 P^0 \tau}\right)\right]}$$

## D. Lancaster

The dynamic equations proposed by Frederick William Lancaster for conflicting populations in 1914 are given by,

$$\frac{dD}{dt} = k_1 \cdot A \cdot D^\gamma \quad , \quad \frac{dA}{dt} = k_2 \cdot A^\beta \cdot D$$

$D$ and $A$ denote the defender and attacker count, respectively.

Dividing both equation sides, $\frac{dD}{dA} = k \cdot A^{1-\beta} \cdot D^{\gamma-1}$, defining $k = \frac{k_1}{k_2}$, separating allows integration of these power functions,

$$\frac{dD}{D^{\gamma-1}} = k \cdot A^{1-\beta} \cdot dA,$$

$$\int_{D_0}^{D} D^{1-\gamma} dD = k \int_{A_0}^{A} A^{1-\beta} \cdot dA, \qquad \frac{D^{2-\gamma}}{2-\gamma} = k \cdot \frac{A^{2-\beta}}{2-\beta}$$

$$D^{2-\gamma} = k \cdot \frac{(2-\gamma)}{2-\beta} \cdot A^{2-\beta} \quad \text{or re-writing} \quad D = \left[ k \cdot \frac{(2-\gamma)}{2-\beta} A^{2-\beta} \right]^{\frac{1}{2-\gamma}}$$

$$D = k^{\frac{1}{2-\gamma}} \left[ \frac{(2-\gamma)}{2-\beta} \right]^{\frac{1}{2-\gamma}} A^{\frac{2-\beta}{2-\gamma}}$$

For the case $\beta = \gamma$

$$D = k^{\frac{1}{2-\gamma}} A \sim k^{0.653595} A$$

Fain re-analyzed data from sixty land battles with quantified judgments of opposing force ratios for an estimated $\beta = \gamma$ which gave a value of 0.47 as the exponent near a square root rational number of 0.5 rather than simpler linear or quadratic values (positive integers of one or two, respectively).

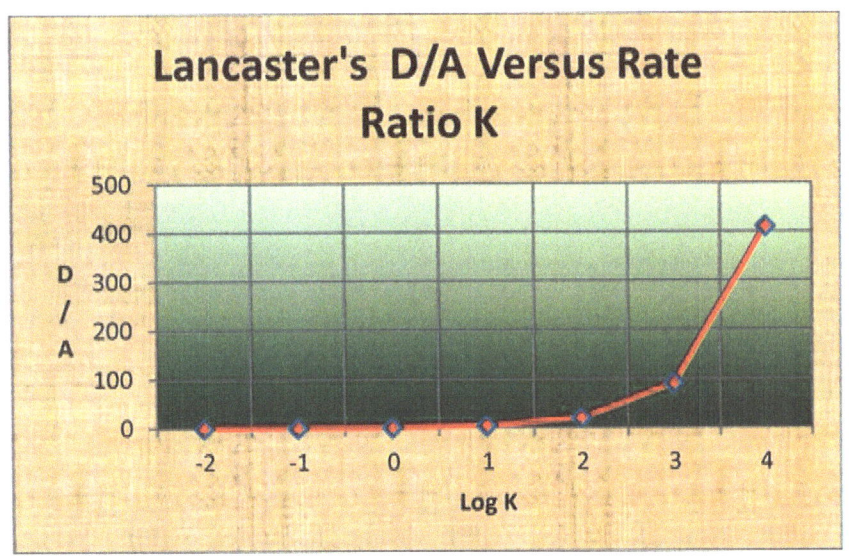

References:

Dupuy, Col. T. N., *Numbers, Predictions and War*, Bobbs Merrill, 1979.

Fain, Janice B., The Lancaster Equations and Historical Warfare: An Analysis of Sixty World War Two land Engagements, Arlington, VA 1975.

Willard, Daniel A., Lancaster as a Force in History: An Analysis of Land Battles in the Years 1618-1905, McLean, VA, 1962.

IIIE. Fractional Derivative

$$\frac{d^{\alpha} P}{dt^{\alpha}} = F(t)$$

$$P - \frac{d^{-\alpha}F}{dt^{-\alpha}} = P - \frac{d^{-\alpha}}{dt^{-\alpha}}\frac{d^{\alpha}P}{dt^{\alpha}} = 0$$

$$P - \frac{d^{-\alpha}}{dt^{-\alpha}}\frac{d^{\alpha}P}{dt^{\alpha}} = c_1 t^{\alpha-1} + c_2 t^{\alpha-2} + \cdots c_m t^{\alpha-m}$$

$$0 < \alpha \leq m < \alpha + 1$$

$$P = \frac{d^{-\alpha}}{dt^{-\alpha}}F(t) + c_1 t^{\alpha-1} + c_2 t^{\alpha-2} + \cdots c_m t^{\alpha-m}$$

For example, let $F(t) = \lambda \cdot t^p$

$$P = \lambda \frac{d^{-\alpha}}{dt^{-\alpha}} t^p + c_1 t^{\alpha-1} + c_2 t^{\alpha-2} + \cdots c_m t^{\alpha-m}$$

For constant $\lambda$, the fractional, q-th differential of a p-th power function is

$$d^q(\lambda \cdot x^p) = \lambda \cdot \frac{\Gamma(p+1)}{\Gamma(p-q+1)} x^{p-q}$$

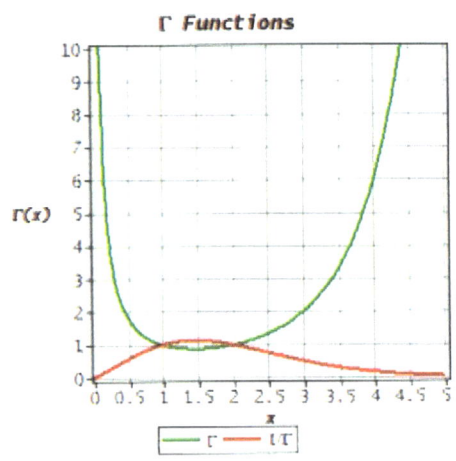

Also, $\quad _c d_t^{-v}(a-t)^{\lambda} = \frac{(a-t)^{\lambda+v}}{\Gamma(v)} B(v, -\lambda - v)$, $\quad v > 0$, $\quad T = \frac{t-c}{a-c}$, $\quad a > t > c \geq 0$

with incomplete Beta function, $\quad B_T = \int_0^T t^{x-1}(1-t)^{y-1} dt \quad 0 < T < 1$

An alternative Laplace Transform technique of solution follows a similar Green's function approach,

$$L\left[\frac{d^{\alpha}P}{dt^{\alpha}}\right] = s^{\alpha}\bar{P} - s^{\alpha-1}\dot{P}(0) - s^{\alpha-2}P''(0)$$

define a Transfer function $K = L^{-1}\left[\frac{1}{s^{\alpha}}\right]$, $L^{-1}\left[\frac{1}{s^{\alpha}}\right] = \frac{t^{\alpha-1}}{\Gamma(\alpha)}$

$$P(t) = \int_0^t K(t-\zeta)F(\zeta)d\zeta + c_1 K + c_2 \acute{K} + \cdots c_N d^{N-1}K$$

IIIF. Biological Predator Prey Population Model

$$\frac{dP_1}{dt} = r \cdot P_1\left(1 - \frac{P_1}{k}\right) - a \cdot P_1 P_2 (1 - e^{b \cdot P_1})$$

$$\frac{dP_2}{dt} = -d \cdot P_2 + e \cdot P_2 (1 - e^{b \cdot P_1})$$

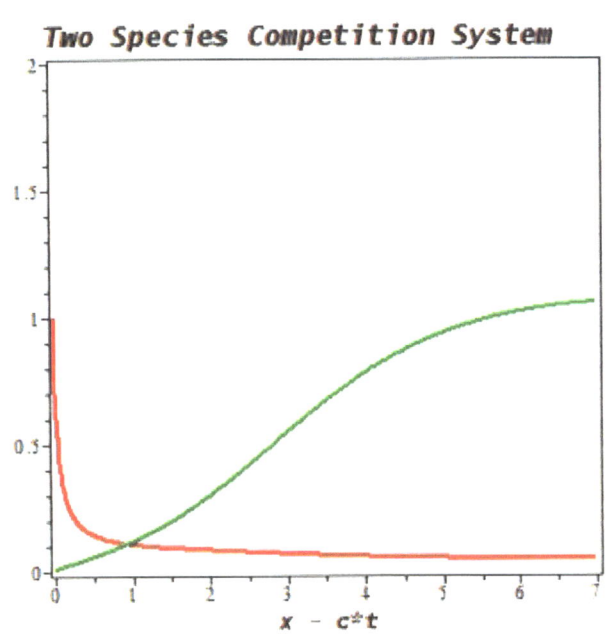

The Voltera and Lotka-Voltera systems have been modified to the following dynamic system.

Consider n individuals of species predator A, m individuals of prey species B, (N-n-m) E which are passive constituents with the following interaction models.

Birth Processes: $BE \to^b BB$

Death Processes: $A \to^{d_1} E \quad B \to^{d_2} E$

Predator-Prey Interactions: $AB \to^{p_1} AA$ , $AB \to^{p_2} AE$

Constant Rates: $b, d_1, d_2, p_1, p_2$

Also, the sum holds $A + B + E = N$

The equations capturing the dynamics are,

$$\frac{dP_1}{dt} = n(P_2)P_1 - \mu P_1$$

$$\frac{dP_2}{dt} = rP_2\left(1 - \frac{P_2}{K}\right) - g(P_2)P_1$$

$\mu = d_1$

$r = 2b - d_2$   The equations form a Lotka-Volterra system when the $\frac{P_2}{K}$ term is missing.

$K = 1 - \frac{d_2}{2b}$,   $n(P_2) = 2P_1P_2$, and

$$g(P_2) = 2(P_1 + P_2 + b)P_2$$

Typical parameter value analyzed are given by,

b=0.5, $d_1 = 0.1$, $d_2 = 0.0$, $p_1 = 0.5$, $p_2 = 0.1$  N=3200

Reference: McKane, A.J. and T.J. Newman, Predator-Prey cycles from resonant amplification of demographic stochasticity, arXiv:q-bio/0501023v1 Jan 2005


## IV.  Multi-Age Mixture Balance Model

Here the continuum balances for the total and partial (by age) populations are derived.

The natural philosophy of the exact sciences has developed a thermodynamic

system model which may be applied to civilizations and institutions.

The continuum mixture system paradigm is summarized by:

- Variables that are multi-variate that are either deterministic or random in a probabilistic framework depending on the model approach
- Equations of state principles : determinism, equi-presence, local and non-local action, transformation or relativity  theory descriptions that imposes a condition on measurements of state variables for example, fluxes of energy or stresses acting on either boundaries or throughout the bodies, be invariant in the defined groups of frames used
- Balances of population number, of attributes or properties, and of energy loss or weakening, thinning dissipative forces or growth and strengthening sources
- An entropy or re-named calory inequality

| Primitives | Mechanics | Thermodynamics |
|---|---|---|
| | Bodies | Single Body |
| | | Mixtures |
| | Time | Time |
| | Place | Place |
| | Mass | Mass |
| | | (Individual and Mixture Total) |
| | | Temperature |
| | | Charge |
| | | Parameters |
| | | Net Working |
| | | Heating |
| | | Internal Energy |
| **General Axioms**: Action-Reaction "Restrictions" | Forces | Thermo-kinetic Interacting Process |
| | Kinetic Processes | |
| | Equilibrium | Stationary State |

A single phase and component equation is derived for a single dimensional system and then made three dimensional. Further a multi-component version is developed to accommodate an age structured population. It is modified for births and deaths of each gender.

The density at a reference time, say t=0 $\rho_0$ is in a defomed space $\rho$ is calculated by use of the mathematical Jacobian $J$. This is referred to as a Lagrangian description of the balance of mass/population.

$$\rho_0 = J\rho,$$

In two dimensions this is a determinant for a ratio of the change of areas in two dimensions and volumes in three.

In a differential element along the x axis, balance the time rate of change within the element and the fluxes through the ends; that is,

$$\frac{\rho(x, t + \Delta t) - \rho(x, t)}{(t + \Delta t) - t} = \frac{\rho(x, t + \Delta t) - \rho(x, t)}{\Delta t} \to \frac{\partial \rho}{\partial t}$$

$$\frac{\rho(x + \Delta x, t)u(x + \Delta x, t) - \rho(x, t)u(x, t)}{(x + \Delta x) - x} = \frac{\partial(\rho u)}{\partial x}$$

$$\frac{\partial \rho}{\partial t} + \frac{\partial (\rho u)}{\partial x} = S$$

The single component balance of mass or population equation in three dimensions obtains in an Eulerian local form,

$$\frac{\partial \rho}{\partial t} + \underline{u} \cdot \nabla \rho + \rho \cdot \nabla \cdot \underline{u} - S = 0 \text{ or}$$

$$\frac{\partial \rho}{\partial t} + \nabla \cdot (\rho \underline{u}) - S = 0$$

For a population mixture of ages $a$, the balance equation is reformulated defining

$\underline{x}_a = \underline{x}_{\kappa,a}(\underline{X}_a, t)$ as reference locations over time, $-\infty < t < \infty$

Also, velocity as a time-differentiation obtains,

$$\underline{\dot{x}}_a = \frac{\partial}{\partial t} \underline{x}_a(\underline{X}_a(\underline{x}, t), t),$$

A practical measure of population age-fraction may be defined, $c_a = \frac{\rho_a}{\rho}$, $c_a(\underline{x}, t)$ as well as the relative velocity $\underline{u}_a = \underline{\dot{x}}_a - \underline{\dot{x}}$. This is historically in physical mechanics is referred to as a diffusion velocity that captures the difference of the individual velocities about the mixture mean velocity. The relative velocity is a difference measure of the total mixture velocity of all the ages and the individual age-group velocity.

Decomposition into male and female components and using summation to match the component age equations to the total population balance equation that holds for all ages obtains,

$$c_a = c_{F,a} + c_{M,a}, \quad c_F = \frac{\rho_F}{\rho} = \sum_{a=1}^{N} c_{Fa}, \quad c_M = \frac{\rho_M}{\rho} = \sum_{a=1}^{N} c_{M,a}, \quad c_{Fa} = \frac{\rho_{F,a}}{\rho}$$

$$c_{M,a} = \frac{\rho_{M,a}}{\rho}, \quad \rho_F = \sum_{a=1}^{N} \rho_{F,a} \quad \rho_M = \sum_{a=1}^{N} \rho_{M,a} \quad \text{Let supply} \quad \hat{c}_a = \frac{S_a}{\rho}$$

$$S = \sum_{a=1,N} S_a$$

The local form of the population balance for each component age is written,

$$\rho \dot{c}_a = -\nabla \cdot (\rho_a \underline{u}_a) + \hat{c}_a$$

$\rho(\underline{x}, t) = \sum_{a=0}^{N} \rho_a(\underline{x}, t)$, Dividing by $\rho$ obtains, $1 = \sum_{a=0}^{N} c_a$.

The net growth at each age $\hat{c}_a = \hat{c}_{F,a} + \hat{c}_{M,a}$ consists of both birth and mortality contributions.

$$\hat{c}_{F,a} = \hat{b}_{F,a} - \hat{d}_{F,a} \cdot c_{F,a}, \text{ and } \hat{c}_{M,a} = \hat{b}_{M,a} - \hat{d}_{M,a} \cdot c_{M,a},$$

Birth moduli for age $a$, account for an age-dependent fertility of the females and males. A partitioning of the births to females at age , $b_{F,a}$ are male. Let the proportion of male births be denoted by $\hat{b}_{M,a} = \mu(a) \cdot b_{F,a}$; the female population by $\hat{b}_{F,a} = w(a) \cdot b_{F,a}$. $\mu(a)$ and $w(a)$ are the fraction parameterizations of the births as males and females.

For independent mortality contributions with probabilities at each age,

$$\hat{d}_{F,a} = 1 - \prod_{i=1}^{N}(1 - p_{i,F,a}), \ \hat{c}_{M,a} = \hat{b}_{M,a} - \hat{d}_{M,a} \cdot c_{M,a}, \ \hat{d}_{M,a} = 1 - \prod_{i=1}^{N}(1 - p_{i,M,a})$$

Additional movements introduced as source or flux terms $J_a$ into or out of the region have been incorporated in previous investigations to account for random movements by classical Laplacian operator diffusion to neighboring lands, or jumps and leaps to distant locales with fractional calculus operators. Furthermore, the diffusive motions are introduced into the continua in different ways. The difference of the total and partial linear momentum equations is a natural alternate. It can lead to a wave diffusive system which has a finite propagation speed for the spreading.

Continuum thermodynamics has enjoyed extensive formalism using an entropy inequality or combining variational calculus to specify the balance equations as constraints on the system understudy and to develop general equations.

## V. Discussion and Conclusion

A quantitative understanding of population modeling using mathematical methods and insights from diverse areas of applied mathematics in biology, physics, and chemistry has been presented. The goals and aims would steer an analysis toward temporal total population, or other alternatives with gender and age dependence are explored here. Analyses and estimations differ with variables defined, for example population number count, or a density on a changing area. A multicomponent continuum model for particles suspended in a fluid using averaged equations has transferable ideas to population dynamics theory. Ideas of multicomponent or multiphase continuum mixtures for age dependence and fractional calculus have been initiated and not previously given in the literature.

## VI. References:

Adams, John A., *Mathematics in Nature: Modeling Patterns in the Natural World,* Princeton University Press 2006.

Arrhenius, S., *Worlds in the Making: The Evolution of the Universe.* New York, Harper & Row, 1908.

Ayati, Bruce P., A Variable Time Step Method For An Age-Dependent Population Model With Nonlinear Diffusion, SIAM J. Numer. Anal., Vol. 37, No. 5, pp. 1571–1589 2000.

Ayati, Bruce P. and Todd F. Dupont, Galerkin Methods in Age and Space For A Population Model With Nonlinear Diffusion, SIAM J. Numer. Anal., 40, 3, 1064-1076 2002.

Bacaer, Nicolas, *A Short History of Mathematical Population Dynamics*, Springer 2011.

Behharbit, Abdelali, T.S. Margulies, and W.H. Schwarz, Finite Amplitude wave propagation through a two-phase system of particles in a viscothermal fluid, J. Acoust. Soc. Am., 91, 5, May pp. 2556-2566, 1992.

Bowen, Ray, *Introduction to Continuum Mechanics for Engineers*: Revised Edition (Dover Civil and Mechanical Engineering), Dover, 2010.

Butzer, Karl W., Collapse, environment, and society, Proc Nat Acad Sci, 2012 March 6, 109(10), 3632-3639.

Dreier, Thomas, *We Human Chemicals*, Updegraff Press, 1948

Halley, Edmond.: An estimate of the degrees of the mortality of mankind, drawn from curious tables of the births and funerals at the city of Breslaw; with an attempt to ascertan the price of annuities upon lives. *Phil. Trans. Roy. Soc. London* **17**, 596–610 (1693).

Heilig, Gerhard K., Anthropologenic Factors in Land-Use Change in China, Population and Development Review, 23, 1 March 1997, pp. 139-168.

Lewellan, Ted C., *The Anthropology of Globalization*, Bergin & Co, 2002.

Liu, I.-S, Method of Lagrange Multipliers for exploitation of the energy principle, Arch. Ration. Mech. Anal., 46, 131-148, 1972.

Margulies, Timothy S., *Location Analysis, Movements, and Renewal*, Create Space Press,

ISBN-13: 978-1477415269


Meindel, Richard S. and Katherine F. Russell, Recent Advances in Method and Theory in Paleodemography, Annual Review of Anthropology, 27, pp. 375-399, 1998

Mueller, I., and Ruggeri, T., *Extended Thermodynamics*, Springer Tracts in Natural Philosophy, 37, Springer, 1993.

Müller, Erich. A. (1998). "Human Societies: a Curious Application of Thermodynamics" *Chemical Engineering Education*, Vol. 1, No. 3, Summer.

Miller, K.S. and B. Ross, *An Introduction to the Fractional Calculus and Fractional Differential Equations*, John Wiley, 1993.

Murray, J.D.,1989. *Mathematical Biology*. Berlin: Springer.

Negrete, D., Fractional diffusion models of non-local transport, Phys. Of Plasmas, 2006.

Oldham, K.B. and Jerome Spanier, 2006, *The Fractional Calculus*, Dover.

Parker, David F., *Fields, Flows, and Waves*, Springer 2003.

Pastor, John, *Mathematical Ecology of Populations and Ecosystems*, Wiley-Blackwell 2008.

Ricker, W.E., Stock and recruitment, J. Fisheries Research Board of Canada, 1954, pp 559-625.

Schwarz, W.H. and T. S. Margulies, Sound Wave Propagation through emulsions, colloids, and suspensions using a generalized Fick's law, J. Acoust. Soc. Am., 90, 6, Dec 1991 pp. 3209-3217.

Sherratt, Jonathan A., Mark Lewis, and Andrew Fowler, Ecological Chaos in the wake of invasion, Proc. Nat. Acad. Sci., 92, 2524-2528, 1995.

Smil, Vaclav, *Energy in Nature and Society: General Energetics of Complex Systems*, MIT Press 2007.

Spooner, Brian J., ed.1972 *Population Growth: Anthropological Implications*. Cambridge: MIT Press.

Truesdell, C.-C Wang, C., *Rational Thermodynamics*, Springer, 1984.

Truesdell, C., *The Elements of Continuum Mechanics*, Springer-Verlag, 1985.

Taylor, L.R. and R.A.J. Taylor, Aggregation, Migration, and Population mechanics, Nature, 265, 1977, 415-421.

Volterra, V., Lecons sur la theorie mathematique de la lute pour la vie, Gauthier-Villars, Paris, 1931.

# Appendix A: Halley's Life Table

Using the City of Breaslau's population data for the years 1687-1691 Edmond Halley formulated the first tabulation for this steady-state system since it assumed no growth as Halley observed the number of births equaled the number of deaths. Define $P_0$ the total annual number of births and $P_k$ the total aged k years, and $M_k$ the number of deaths at age k.

He computed $P_{k+1} = P_k - M_k$ or $M_k = P_k - P_{k+1}$ for $k \geq 0$.

Also, $\frac{P_{k+1}}{P_k}$ is the relative probability of surviving until age $k + 1$.

Another problem addressed by Halley was calculating life annuities. The problem was addressed earlier by De Witt Prime Minister of Holland in 1671; however, the method was lost as Holland was invaded and De Witt lynched.

Let price $A_k = \frac{1}{P_k}\left(\frac{P_{k+1}}{1+r} + \frac{P_{k+2}}{(1+r)^2} + \frac{P_{k+3}}{(1+r)^3} + \cdots\right) = \frac{1}{P_k}\sum_{i=1}^{N}\frac{P_{k+i}}{(1+r)^i}$ A person has probability of survival at age k+N of $\frac{P_{k+N}}{P_k}$. The factor $1/A_k$ obtains the desired annuity. The example from

$A_k = 12.27$ using Halley's assumed interest rate of 6% obtaining $\frac{1}{12.27} \sim 0.085$ or 8.5 % of the initial sum each year.

# Appendix B: Survival Probability

The probability $S(t)$ that a specified event has not occurred in a system for a given period of time t and under specified conditions. The failed system state is denoted by $F(t)$ where $S(t) = 1 - F(t)$, $P(t) \geq 0$

Using lower case letters for the probability density functions and capital letters for the cumulative distribution function,

$$\int_t^\infty f(x)dx \qquad f(x) \cdot dx = -dS$$
where

A function called a hazard function is defined by,

$$\lambda(t) = \frac{f(t)}{S(t)}, \text{ where } F(t) = 1 - S(t), \text{ so that } \frac{d}{dt}F(t) = f(t) = -\frac{dS}{dt}$$

Note $\lambda(t) = \frac{dS}{(1-F(t))dt} = -\frac{d}{dt}\ln(1 - F(t))$ and

$$\lambda(t) = \frac{-d}{dt}\ln[S(t)] = \frac{-1}{S(t)}\frac{dS}{dt} = \frac{f(t)}{S(t)}$$

And finally $S(t) = \exp\left(-\int_0^t \lambda(\sigma)d\sigma\right)$ with mean life-time $M = \int_0^\infty t \cdot f(t) \cdot dt$

Using an Exponential probability density as an example,

$$f(t) = \lambda \cdot \exp(-\lambda t) \quad S(t) = \exp\left(-\int_0^t \lambda(\sigma)d\sigma\right) = \exp(-\lambda t)$$

$$M(t) = \int_0^\infty t \cdot \lambda \exp(-\lambda t)dt \quad M(t) = \frac{1}{\lambda}$$

The ideas of estimating a population survival by a life-table by a simple difference quotient calculational approach follows for a first derivative model.

$$\frac{dF}{dt} \approx \frac{F(t+\Delta t) - F(t)}{\Delta t}$$

$$\frac{n(t+\Delta t) - n(t)}{N \cdot \Delta t} \text{ for total count n as N}$$

$$\lambda(t) = \frac{f(t)}{R(t)}$$

Survival Probability Life Table

| Age (Yrs) | # Living L(t) | R(t) = L/N | F(t) = 1 - R(t) | f(t) | λ(t) |
|---|---|---|---|---|---|
| 0 | 100 | 1.00 | 0 | 0.02 | - |
| 1 | 98 | 0.98 | 0.02 | 0.01 | 1/98 |
| 2 | 97 | 0.97 | 0.03 | 0.03 | 3/97 |
| 3 | 94 | 0.94 | 0.06 | | |
| | 0 | 0 | 1.0 | | |

The intent was to be instructive with a simple example. Estimators for the probabilities for example of a Weibull model (such as Kaplan Meier or other maximum likelihood or entropy estimators) and numerical derivatives such as that derived by the Richardson formula to reduce error are chosen in application. Disruptive events ranging from cracks in metals, to earthquakes, as well as fractures in society have been idealized and modeled with hazard functions and survival probability calculations.

References:

Weibull, Walodi, *A Statistical Distribution of Wide Applicability*, ASME J. Applied Mech., September 1951, pp. 293-297.

Stevenson, Mark, *An Introduction to Survival Analysis*, EpiCentre, IVABS, Massey University, June 4, 2009.

## Appendix C: Perspectives on the Transformation or Decline of a Civilization

Societies are born, grow and decline as stages of a life cycle. The use of human beings and the lands of the Earth takes many forms. In particular, a broad perspective on land use evolution is attempted in the table below.

**Land Use Changes**

| Global Driving Forces | Proximate Determinants of Land Use-use Change |
|---|---|
| Spread of Scientific Methods and Advanced Technology | Rapid Expansion of Transportation & Communication Networks |
| Geo-political & Economic Structures and Strategies | Increases in Mass-mobility and Tourism |
| Changes in Life-Styles | Modernization & Expansion of agriculture and livestocks |
| Population Growth | Growing Demand for Commercial Energy |

| Land-Cover Modifications |
|---|
| Deforestation & Forest Modification |
| Draining of Wetlands |
| Dam Construction, River Regulation |
| Man-made Desertification |
| Soil Erosion, Land Slides |
| Chemical & Nuclear Contamination of Soils & Water Bodies |
| Land Sealing (buildings, streets) |

Ref: [Heilig]

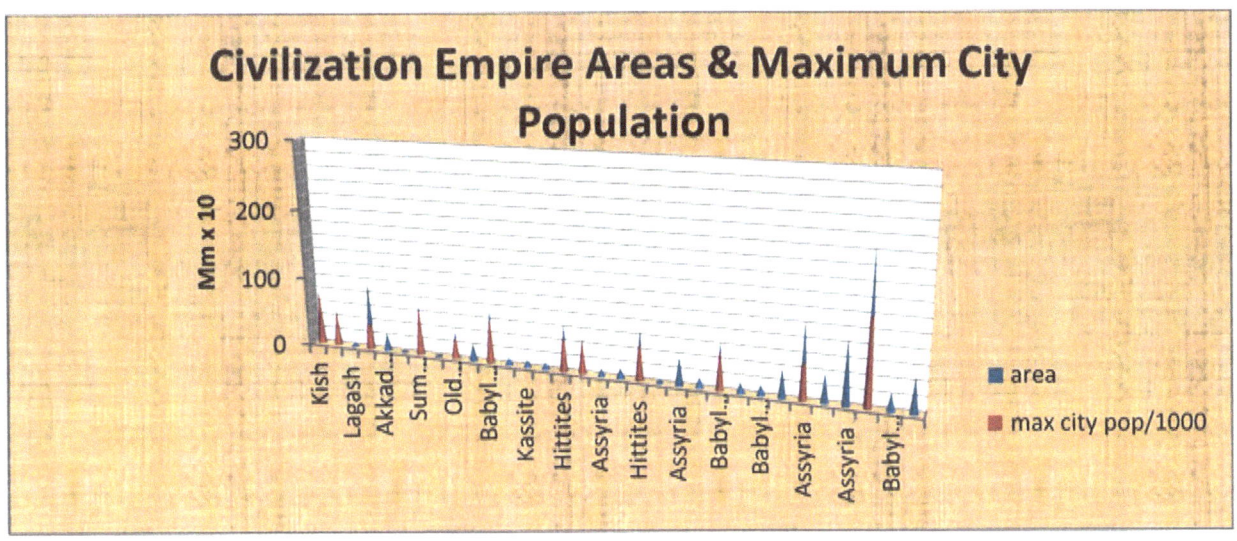

Data from Chase-Dunn, Chris, Alvarez, Alexis, and Pasciuti, Thomas Hall [Univ. CA Report]

The End of the Western Roman Empire has been the traditional marker for entering the Middle Ages. In Carroll Quigley's, *The Evolution of Civilizations*, as a society transforms from an instrument of expansion to only self-serving needs it becomes evitable to decay. He depicts seven stages of a civilization: mixture, gestation, expansion, conflict, universal empire decay, and invasion. Capitalism was an instrument of expansion with a central banking system as necessary for prosperity.

Historians theorize on understanding the similarities and differences that surround the continued existence of the Eastern or Byzantine Empire which survived about a thousand years after the fall of the Western. Gibbons implicates Christianity in the Western Empire's fall, yet the Eastern was more Christian in numbers and geographic extent. Environmental weather factors appear approximately the same for both Eastern and Western regions; however, the East did not fall.

Revenge trumps other inhibitory signals for engaging in conflicts especially when activated by hatred. For example, Charles V sent troops to invade Rome motivated by the Pope (Clement VII) joining the anti-Imperial league in 1527. The troops were comprised mostly of Lutheran mercenaries intent to revenge their persecution by the Catholic Church. The engagement devastated the minds of intellectuals and artists who decided to disperse geographically as their hope of fortune must be found elsewhere.

An alternate view of a disastrous "fall" is that there was "no fall" but a series of events and modifications as displayed in any natural system's dynamics.

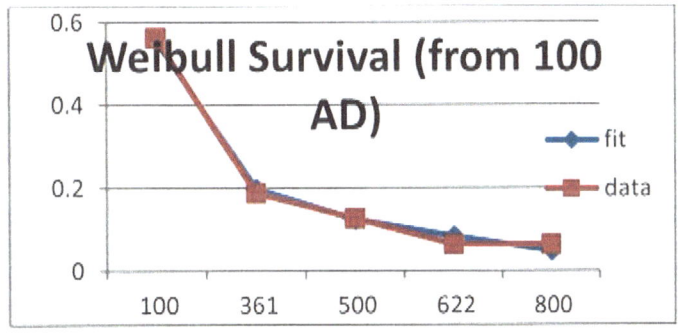

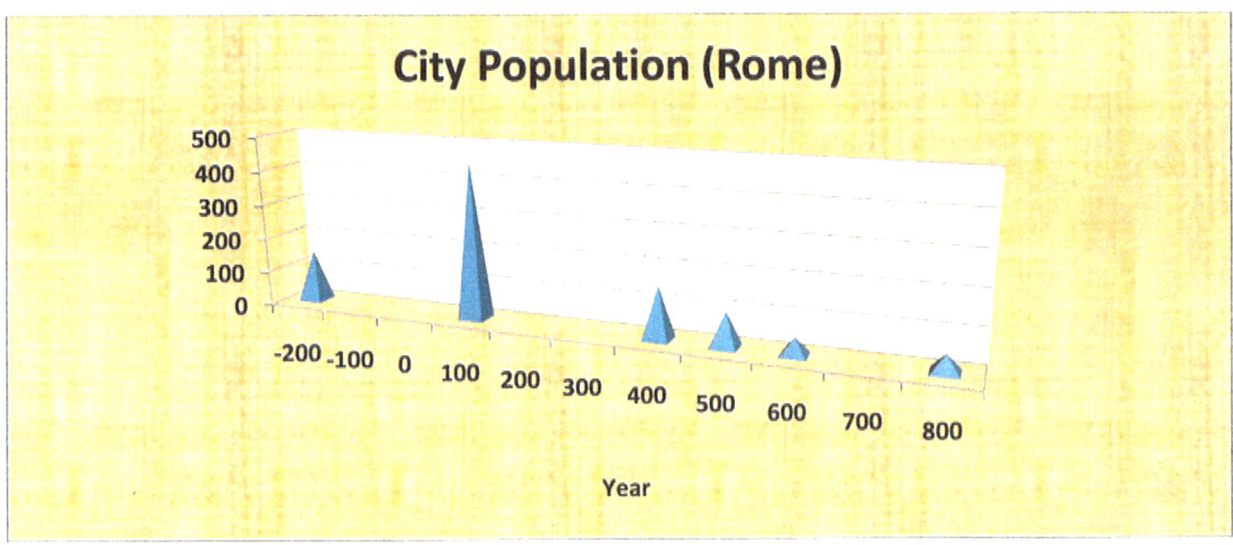

Fitting to a Weibull $P(t) = e^{-a \cdot t^\beta}$ by a logarithmic transformation obtains a linear equation for regression. A fit was made with $R^2 = 0.98$.

An organization with successful management would succeed and not die. An organization's growth may not be synonymous with typical ideas of expansion. A theory for civilization's movements and stages, economies, or cycles may be developed along this viewpoint. Effective management must meet head-on its challenges, promote communication, be sensitive to problems to instill trust and confidence in the organization. Downsizing may be implemented as part of the evolution. As employees have trouble accepting the closing, it has been found that neither productivity plummeted along the final stage nor sabotage increased.

A learning organization continually enhances its capacity to create and to compete. Learning disabilities may appear as when people form a strong identification with their position and roles without seeing how they fit within the larger organization. This contributes to animosity towards others, especially as something goes wrong. Experiential learning with feedback could benefit the organization. However, management may be viewed as "skilled but incompetent" to get the jobs done. They keep themselves safe from threats or embarrassments, and thus pursue a path of minimal learning to really help the organization. Important factors for a learning organization are: building a shared vision held by individuals, demonstrating a commitment to the vision, constructing a mental model of reality internally to help balance advocacy and inquiry, recognizing that shared mental models, "of the same mind", contribute to organizational learning, and submitting to a systems approach of monitoring, providing feedback, and implementing changes.

The right aim and vision would entail centering a learning organization on community characterized by a lifelong learning process in a complex society, by responsiveness to needs and by making changes while conscious of alternative courses of actions. Communities provide an environment of safety conducive to good communications and to balanced growth.

References:

Chris Chase-Dunn, Alexis Alverez, Dan Pasciuti, and Thomas Hall, Power and Size: Urbanization and Empire Formation in World Systems, UC Riverside, 2002.

Chase-Dunn, Chris And Susan Manning, City systems and world-systems: Four millennia of city growth and decline, The Institute for Research on World Systems, UC Riverside 2002.

Weibull, Waloddi, A Stistical Distribution of Wide Applicability, ASME J. Applied Mech., pp 293-297, September 1951.

# Appendix D: Continuum Area Average Population Model

Let the total cohort population $P_a$ of age $a$ in an area that may change be calculated as the sum,

$P_a = \rho_a A + M_a + S_a$, at time t = 0, $P_{a0} = \rho_{a0} A_{a0} + M_{a0} + S_{a0}$ where the functions for total population, area density population, area rate of change, and net migration areal density are given by $P_a(x,t), \rho_a(x,t), A(x,t),$ and $M_a(x,t)$ respectively with specification of its spatial-temporal dependence. $S_a$ denotes other contributions to growth or decline.

$$\dot{P}_a = \overline{(\rho_a A)} + \dot{M}_a + \dot{S}_a,$$

Differentiating the product obtains, $\dot{\rho}_a A + \rho_a \dot{A} + \dot{M}_a + \dot{S}_a = \dot{P}_a$ where the total population in the defined area is given by $P = \sum P_a$ and $\sum \dot{P}_a = \dot{P}$.

The theoretical dynamics of physical systems is comprised of dynamical equations for linear momentum, angular momentum, and energy or entropy. Here a overview for a single individual or age is given. That is, movements are defined by $\underline{m} = \int_B \underline{\dot{x}}\, dP$ with rate of change force balance equation $\underline{\dot{m}} = \underline{F}$ where the forces are decomposed into surface or boundary contact and body forces that act throughout the body.

$\underline{F} = F_C + F_B$   The science of mechanics also defines for rotational motions an angular movement version of these natural body motions by

$\underline{\dot{A}} = \Im$   total torque $\Im$   where   $\underline{A} = \int_B (\underline{x} - \underline{x}_0) \wedge \underline{\dot{x}}\, dP$

The total energy E written as a summation of kinetic T, internal I, along with a source term $\Psi$ may be written as,

$$E = T + I + \Psi \qquad \text{or re-writing with} \qquad \Psi = 0$$

$E - T = I$.   Obtaining   $\delta(E-T) = \delta I$,

By the First Law of Thermodynamics, the change in the internal energy $\delta I$ of the sub-system is the sum of the heat $\delta Q$ added, and net work $\delta W$.

$$\delta I = \delta W + \delta Q$$ In terms of specific quantities, $\varepsilon = \dfrac{I}{P}$, $\delta I = \int \rho \varepsilon \cdot dP$

When there is no heating $\delta Q = 0$. The entropy or calory $\delta C$ as renamed by Truesdell

$$\delta C = \int \eta \cdot dP$$ with specific, per unit mass caloric $\eta$ at temperature $\theta > 0$.

Changes of Internal Energy by chemical reactions would be written in terms of chemical potentials $\mu_i$ and mole changes. $N_i$  $\quad \delta I_{React} = d\left( \sum_i \mu_i N_i \right)$

Considering rates of change $\quad \overline{\delta I}' = \overline{\delta Q}' + \overline{\delta W}'$

When $\overline{\delta Q}' = 0$, $\int \rho \varepsilon \cdot dP = \delta W$, and calory cannot decrease $\delta C' \geq 0$ or

$\int \eta' \cdot dP \geq 0$. Entropology is a term introduced for application of the ideas of calory or entropy to societal dynamics. Energy supplies are needed to sustain a more orderly state as disorder accompanies the calory increase if not countered as in a statistical interpretation.

The term *anthropology* is from the Greek *anthrōpos* (άνθρωπος), man meaning mankind or humanity, and *-logia* (-λογία), meaning study. The German philosopher Magnus Hundt is credited with the term's first usage. The term entropy was coined in 1865 by Rudolf Clausius based on the Greek εντροπία [entropía], a turning toward.

References:

Bailey, Kenneth, *Social Entropy Theory*, State University of New York, 1990.

## Appendix E: Statistical Entropy Interpretations

Shannon information entropy of a system (as a thermodynamic one) where entropy is a measure of randomness, smoothness, or disorderliness attempts to quantify the transfers of matter-energy or communication information.

A statistical interpretation of entropy may be developed from taking into account several ideas including: the increase in microstates increases entropy, as well as 2) the microstates of a combined state of two separate systems is the product of the microstates; however, the entropy of the combined state is the sum (result of addition) of entropies. The logarithm function ( a mathematical inverse function of the exponential function) has properties that may be used in constructing a mathematical representation of this combined state.

Empirical Probability $\quad P(E) = \dfrac{frequency\ of\ event\ E}{number\ of\ trials\ of\ experiment} \quad 0 \leq P(E) \leq 1$

Classical Probability $\quad P(E) = \dfrac{number\ of\ ways\ E\ can\ occur}{number\ of\ possible\ experiment}$

Let the entropies of the two states A and B be denoted by $E_A$ and $E_B$, where $E_A + E_B = E_{AB}$ is the combined state.

$E_A = k \cdot log(W_A), E_B = k \cdot log(W_B)$

$E_{AB} = k \cdot log(W_A) + k \cdot log(W_B) = k \cdot log(W_A \cdot W_B)$

For equi-probable microstates, $p_i = \dfrac{1}{W_{AB}} \quad$ or $\quad W_{AB} = \dfrac{1}{p_i}$

For calory $C = k \cdot log(W_A \cdot W_B)$, $C = k \cdot log\left(\dfrac{1}{p_i}\right)$ or using properties of logarithms

$C = k \cdot log(1) - k \cdot log(p_i)$ and since the logarithm of one is zero, $C = -k \cdot log(p_i)$.

The expectation is the summation for this discrete system $C = \sum p_i x_i = -k \sum_i p_i log(p_i)$

Here $x_i = -k \cdot log(p_i)$ which may be referred to as a Boltzmann variable. This formula is attributed to Shannon with application to communication systems and information storage and transfer. The maximum entropy is obtained for the system with equi-probable microstates or events.

Consider a chemical biological system such as the base compositions of the molecule DNA (deoxyribonucleic acid) whose structure was illuminated by Watson and Crick. The set of four bases $\{A, T, C, G\}$ compose different arrangement on the strung of the ladder or helical strand where

$A = Adenine$, $T = Thymine$, $C = Cytosine$, and $G = Guanine$

One would estimate the entropy of the molecular system by $C = -k \sum p_i \log(p_i)$ from estimates of the probabilities $p_i$. When the probabilities of the bases are equal, equi-probable $p_i = \frac{1}{4}$, then the entropy state is a maximum. This may be the case for the DNA of Escherichia coli micro-organisms. The entropy calculation may be easier from data on conditional probabilities.

| Lower Entropy | Higher Entropy |
|---|---|
| Non-random | Random |
| Organized | Disorganized |
| Ordered | Disordered |
| Separated | Mixed |
| Not equi-probable | Equi-probable |
| Dependent events | Independent events |
| Arrangements restricted | Configuration variety |
| Constraint | Freedom (of Choice) |
| Less uncertainty (more reliable) | Uncertainty |
| Fidelity | Higher error probability |

Distribution of energy among sites of people may be estimated from a statistical physics viewpoint. A mixture of distributions is modeled with this extension of the classical statistical thermodynamic distributions as special cases for Bosons which satisfy the Bose-Einstein distribution and Fermions which accommodate the Fermi-Dirac distribution in statistical thermodynamics [Wu]. An individual or particle quantum state may consist of an arbitrary number of identical bosons, while no two identical fermions can occupy one and the same quantum state as a Pauli exclusion principle.

Calory $C = (E - \sum_i \mu_i N_i - \Omega)/\theta$ Thermodynamic temperature $\theta$ and potential $\Omega$ written in terms of a partition function Z : $\Omega = -k\theta \ln Z$, total energy $E = \sum_i N_i \epsilon_i$

$$C = k \sum_i G_i \left\{ n_i \frac{\epsilon_i - \mu_i}{k\theta} + \ln \frac{1 + n_i - \sum_j \beta_{ij} n_j}{1 - \sum_j \beta_{ij} n_j} \right\}$$

The number of independent states $G_i$ of a single particle species I with energy $\epsilon_i$

$$W = \frac{[G+(N-1)(1-\alpha)]!}{N!(G-\alpha N-(1-\alpha))!} \qquad 0 < \alpha < 1$$

where $W_b = \frac{(G+N-1)!}{N!(G-1)!}$; $W_f = \frac{G!}{N!(G-N)!}$ or more generally,

$$W = \prod_i \frac{[G_i+N-1-\sum^j \alpha_{ij}(N_j-\delta_{ij})]!}{N_i!(G_i-1-\sum^j \alpha_{ij}(N_j-\delta_{ij}))!}$$

This accounting captures local particle states' change as particles are added with fixed size and boundary conditions; however, statistical interactions $\alpha_{ij}$ are considered linear. For ideal gases $\alpha_{ij} = \alpha \delta_{ij}$; I is a single species energy level with $\alpha = 0,1$ for classical Bose and Fermi cases.

References:

Gatlin, Lila L., *Information theory and living systems*, Columbia Univ. Press, 1973.

Kittel, Charles and Herbert Kroemer, *Thermal Physics*, WH Freeman 1980.

Prigogine, I. and R. Defay, *Chemical Thermodynamics*, Translated by D.H. Everett, (Longman, 1954).

Shannon, Claude E. and Warren Weaver, 1949, *The Mathematical Theory of Communication*, Univ. Illinois Press.

Wu, Yong-Shi, Statistical Distribution for Generalized Ideal gas of Fractional Statistics Particles, Phys. Rev. Let, 73, 7, 15 August 1994, 922-925.

## Appendix F: Lattice Models

The Hamiltonian development of dynamics enjoys generalized coordinates, multi-variable calculus, geometric interpretations, as well as a simplicity in the final equations' notation. Furthermore, with restrictions to potential energy independence of $q_j$ and $\frac{\partial H}{\partial t} = \frac{\partial L}{\partial t} = 0$, The H function is a constant equivalent to the sum of kinetic and potential energies. The early Toda lattice which was believed relevant as a regular array with exponential influences among the sites with modifications made for actions-reactions. In a dynamical system with a periodic orbit and initial conditions, generating maps of the flows from the point at which this orbit first returns, forms a first recurrence map named for Henri Poincare and re-named as a first recurrence map. These maps complement scatter plots of time- and frequency-domain analysis.

$H = \sum_{j=1}^{N} p_j \dot{q}_j - L(q_j, \dot{q}_j, t)$   Define generalized momentum $p_j = \frac{\partial L}{\partial \dot{q}_j}$   $j = 1..N$ and Lagrangian function $L$, and location coordinates $q_j$

$$dH = \sum_j \dot{q}_j \, dp_j + \sum_j p_j \, d\dot{q}_j - dL$$

Differentiating and using the Lagrangian from classical mathematical physics obtains,

$$H(q_j, \dot{q}_j, t) \quad, \quad dH = \sum_j \frac{\partial H}{\partial q_j} dq_j + \sum_j \frac{\partial H}{\partial \dot{q}_j} d\dot{q}_j + \frac{\partial H}{\partial t} dt$$

$$dH = \sum_j \dot{q}_j \, dq_j + \sum_j p_j \, d\dot{q}_j - \sum_j \frac{\partial L}{\partial q_j} dq_j - \sum_j \frac{\partial L}{\partial \dot{q}_j} d\dot{q}_j - \frac{\partial L}{\partial t} dt$$

$$\frac{d}{dt}\frac{\partial L}{\partial \dot{q}_j} - \frac{\partial L}{\partial q_j} = Q \quad, \text{with irreversible contribution } Q \text{ and } \dot{p}_j = \frac{\partial L}{\partial q_j}$$

$$\dot{q}_j = \frac{\partial H}{\partial p_j}, \quad -\dot{p}_j = \frac{\partial H}{\partial q_j} \qquad H = \sum_j \dot{q}_j \, dq_j - \sum_j \dot{p}_j \, d\dot{q}_j - \frac{\partial L}{\partial t} dt + Q$$

A selected form of the Hamiltonian was used in the calculations shown.

$$H = 0.5(p_1^2 + p_2^2) + \frac{1}{24} e^{2q_2 + 2\sqrt{3} q_1} + \frac{1}{24} e^{2q_2 - 2\sqrt{3} q_1} + \frac{1}{24} e^{-4q_2} + \frac{1}{8} q_1^2 q_2 - 3 q_1^3$$

Initial conditions: t = 0; $p_1 = 0.1, p_{12} = 1.4, q_1 = 0.1, q_2 = 0$

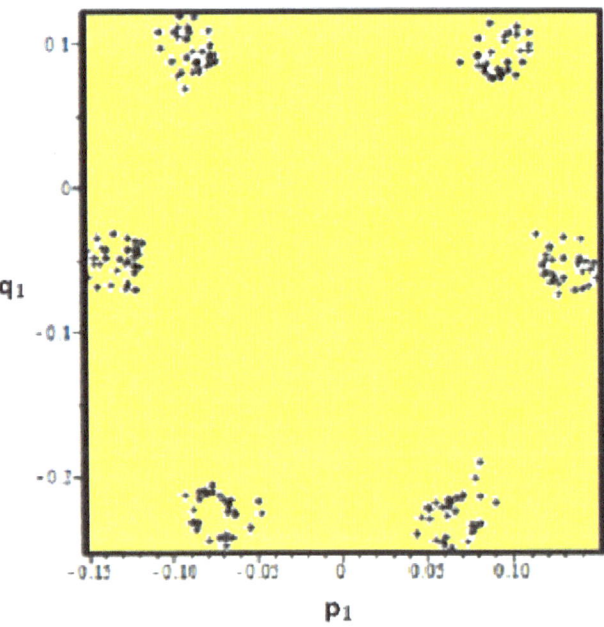

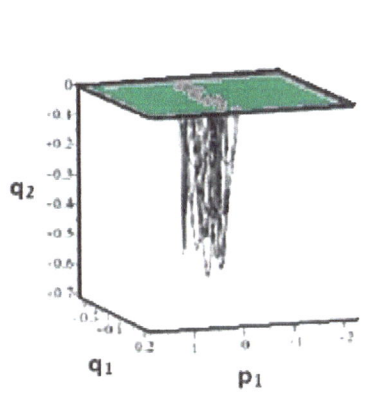
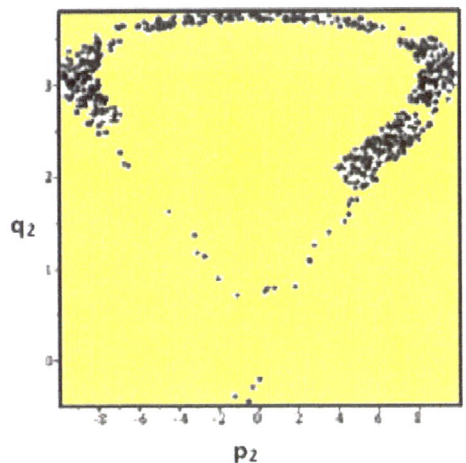

Toda, M., Prog. Theor. Phys., Suppl., 45, 174, 1970.

Landau, L.D. and E.M. Lifshitz, *Mechanics*, Third Edition: Volume 1 (Course of Theoretical Physics) Butterworth-Heinemann; 3 edition 1976.

# Appendix G: Hurst Exponent Time Series Analysis

The Hurst exponent for a time series $X_1, X_2, X_3, \dots X_n$,

is calculated as follows. It gave interesting results for a variety of analyses originating from data analysis of the Nile River flooding region.

1) Calculate a mean value $M = \frac{1}{n}\sum_{i=1}^{n} X_i$,

2) Then adjust the series by subtracting the mean to obtain a mean of zero,

   $Y_t = X_t - M$ for $t = 1,2..n$

3) Form a cumulative series, $Z_t = \frac{1}{n}\sum_{i=1}^{n} Y_i$,

4) Calculate a range series, $R_t = \max\{Z_1, Z_2..Z_t\} - \min\{Z_1, Z_2..Z_t\}$, $t = 1,2..n$

5) A standard deviation is calculated by, $S_t = \sqrt{\frac{1}{t}\sum_{i=1}^{t}(X_i - m_t)^2}$   $t = 1,2..n$ where $m_t$ is the mean value calculated from $X_1$ to $X_t$

6) A rescaled range is formed by $\frac{R_t}{S_t}$ or renaming $\left(\frac{R}{S}\right)_t$ for $t = 1,2..n$

Then a fit is made $\left(\frac{R}{S}\right)_t = c \cdot e^H$ by plots or regression. The analysis for the Nile River gave H=0.91 greater than 0.5 which was interpreted as a non-random Brownian motion process.

Hurst, H.E., The Long-term Storage Capacity of Reservoirs, Trans. Am. Soc. Civil Engineers, 116, 1951.

## Appendix II: Fractional Calculus Applied to Positive Continuous Time Linear Systems

System state space approach for a fractional calculus positive continuous time linear system with reachability,

$$d^\alpha x(t) = Ax(t) + Bu(t)$$

$$y(t) = Cx(t) + Du(t)$$

$x$  System State

$y$  System Output

$u$  System Input

$$x(t) = \Phi_0 x_0 + \int_0^t \Phi(t-\tau) Bu(\tau) d\tau, \quad x(0) = x_0$$

$$\Phi_0(t) = E_\alpha(At^\alpha) = \sum_{k=0}^\infty \frac{A^k t^{k\alpha}}{\Gamma(k\alpha+1)}, \quad \Phi(t) = \sum_{k=0}^\infty \frac{A^k t^{(k+1)\alpha-1}}{\Gamma((k+1)\alpha)}$$

$$L[d^\alpha x(t)] = s^\alpha X(s) - s^{\alpha-1} x_0, \quad X(s) = L[x(t)] = \int_0^\infty x(t) e^{-st} dt$$

$$X(s) = [I_N s^\alpha - A]^{-1}(s^{\alpha-1} x_0 + BU(s)), \quad U(s) = L[u(t)]$$

$$[I_N s^\alpha - A]^{-1} = \sum_{k=0}^\infty A^k s^{-(k+1)\alpha}$$

$$[I_N s^\alpha - A]\sum_{k=0}^\infty A^k s^{-(k+1)\alpha} = I_N$$

$$X(s) = \sum_{k=0}^\infty A^k s^{-(k+1)\alpha} x_0 + \sum_{k=0}^\infty A^k s^{-(k+1)\alpha} BU(s)$$

$$x(t) = L^{-1}[X(s)] = \sum_{k=0}^\infty A^k L^{-1}[s^{-(k+1)\alpha}] x_0 + \sum_{k=0}^\infty A^k L^{-1}[s^{-(k+1)\alpha} BU(s)]$$

$$x(t) = \Phi_0(t) x_0 + \int_0^t \Phi(t-\tau) Bu(\tau) d\tau$$

$$\Phi_0(t) = \sum_{k=0}^{\infty} A^k L^{-1}\left[s^{-(k\alpha+1)}\right] = \sum_{k=0}^{\infty} \frac{A^k t^{k\alpha}}{\Gamma(k\alpha+1)}$$

$$\Phi(t) = L^{-1}\left[I_N s^\alpha - A\right]^{-1} = \sum_{k=0}^{\infty} A^k L^{-1}\left[s^{-(k+1)\alpha}\right] = \sum_{k=0}^{\infty} \frac{A^k t^{(k+1)\alpha-1}}{\Gamma((k+1)\alpha)}$$

Example with Maple Calculation follows.

$$\alpha := 0.5$$

$$A := \begin{bmatrix} 0 & 1 \\ 0 & 0 \end{bmatrix}$$

$$B := \begin{bmatrix} 0 \\ 1 \end{bmatrix}$$

$$x0 := \begin{bmatrix} 1 \\ 1 \end{bmatrix}$$

$$\begin{bmatrix} x1(t) \\ x2(t) \end{bmatrix} = \begin{bmatrix} 1 + 1.128379167 t^{0.5} + 1. t^{1.0} \\ 1 + 1.128379167 t^{0.5} \end{bmatrix}$$

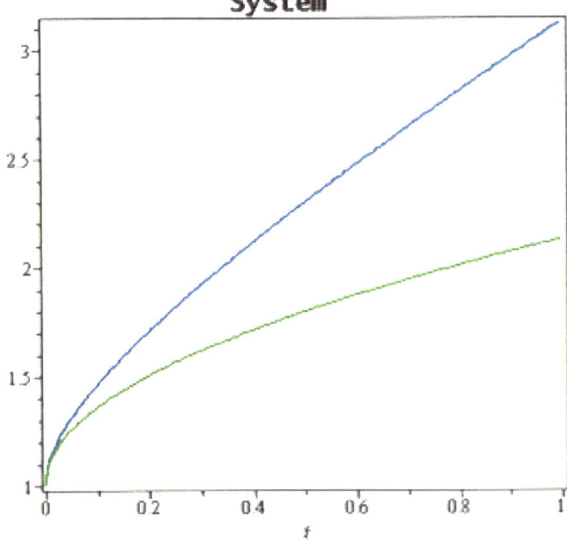

**Fractional Calculus Kalman Filter System**

Kaczorek, Tadeusz, Fractional Positive Continuous Time Linear Systems and Their Reachability, Int. J. Appl. Math. Comput. Sci., 2008, Vol. 18 No. 2, 223-228.

## Appendix I: Continuum Transport Balance Equation

For area regions $A_i$, a continuous density function $\rho_i = \frac{\Delta P}{\Delta A_i}$ where $\rho(\underline{x}(\underline{X},t)) = \lim_{\Delta A_i \to 0} \rho_i$

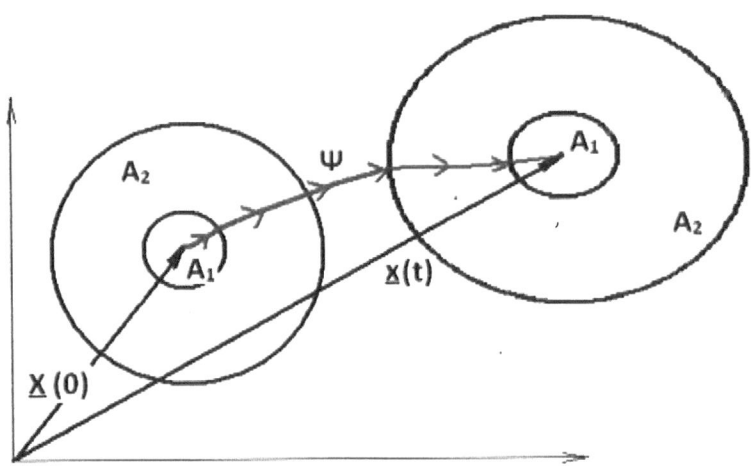

$\Psi$ represents a mapping of the motion starting at t=0 with an example space point $\underline{X}(0)$ and at a later time $\underline{x}(t)$. The motion may be described by an Eulerian system in space, say of one dimension with partial derivative

$\frac{\partial \Psi}{\partial t}(x,t) = \frac{\Psi(x,t+\Delta t) - \Psi(x,t)}{\Delta t}$ whereas the Lagrangian derivative which follows the particle motion is

written $\dot{\Psi}(X,t) = \frac{d\Psi}{dt} = \frac{\Psi(X,t+\Delta t) - \Psi(X,t)}{\Delta t}$

Velocity $v = \dot{\Psi}(X,t)$ and $x(X,t) = X + \int_0^t v(X,\tau)\,d\tau$ Let subscripts be used to help denote the different Eulerian and Lagrangian descriptions, as well as to relate their differential in time, $\Psi_E(x,t) = \Psi_E(x(X,t),t) = \Psi_L(X,t)$

$\dot{\Psi}_L(X,t) = \frac{\partial \Psi_E(x,t)}{\partial t} + \frac{\partial \Psi_E(x,t)}{\partial x}\dot{x}$ and $v = \dot{x}$.

For the population $P = \int_{a(t)}^{b(t)} \rho(x,t)\,dx$ the change is calculated by

$$\frac{dP}{dt} = \frac{d}{dt}\int_{a(t)}^{b(t)} \rho(x,t)\,dx = \int_{a(t)}^{b(t)} \frac{\partial \rho(x,t)}{\partial t}\,dx + \rho(b,t)\frac{db}{dt} - \rho(a,t)\frac{da}{dt}$$

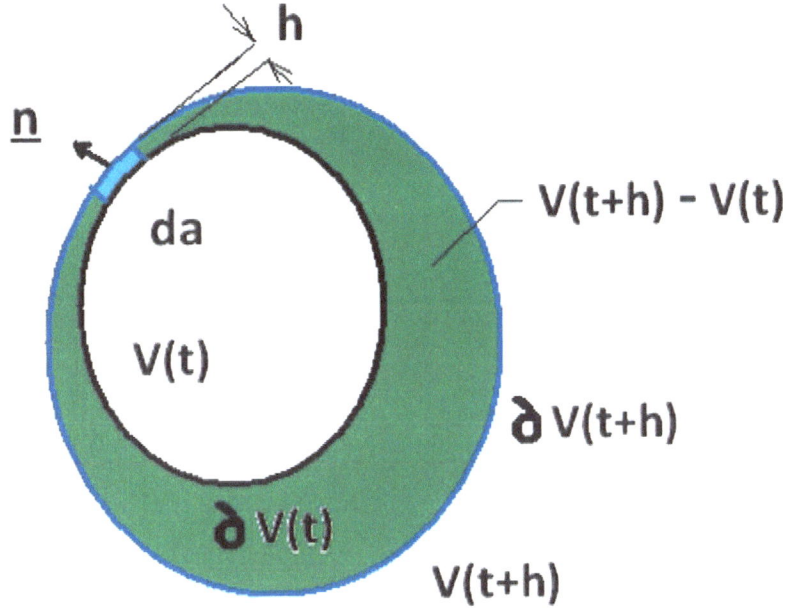

$$\frac{d}{dt}\int_V \Phi dV = limit_{h\to 0}\frac{1}{h}\left\{\int_{V(t+h)} \Phi(x,t+h)dV - \int_{V(t)} \Phi(x,t)dV\right\}$$

$$= limit_{h\to 0}\frac{1}{h}\left\{\int_{V(t+h)} \Phi(x,t+h)dV - \int_{V(t)} \Phi(x,t+h)dV\right\}$$

$$+ limit_{h\to 0}\frac{1}{h}\left\{\int_{V(t)} \Phi(x,t+h)dV - \int_{V(t)} \Phi(x,t)dV\right\}$$

$$= limit_{h\to 0}\frac{1}{h}\left\{\int_{V(t+h)-V(t)} \Phi(x,t+h)dV - \int_{V(t)} \frac{\partial}{\partial t}\Phi(x,t)dV\right\}$$

$$= limit_{h\to 0}\frac{1}{h}\int_{V(t)} \Phi(x,t+h)dV = limit_{h\to 0}\frac{1}{h}\left\{\int_{\partial V(t)} \Phi(x,t+h)u_n h\, da\right\}$$

$$= \left\{\int_{V(t)} \Phi(x,t+h)u_n da\right\}$$

For a material region, $u_n = \underline{\dot{x}} \cdot \underline{n}$

$$\frac{d}{dt}\int_V \Phi dV = limit_{h\to 0}\frac{1}{h}\int_{V(t)} \Phi(x,t+h)dV = \left\{\int_{V(t)} \Phi(x,t+h)\underline{\dot{x}}\cdot \underline{n}da\right\}$$

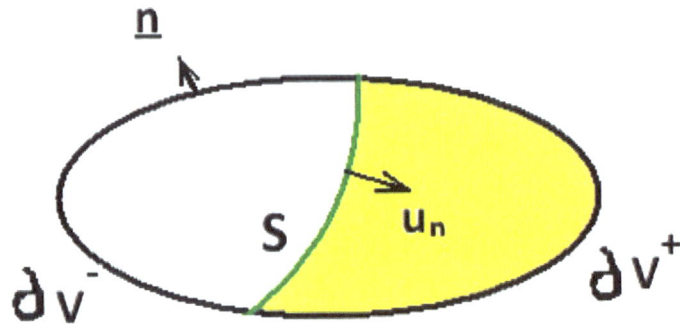

$u_n$ is the normal speed of the singular surface S pointing in the direction of $\underline{n}^+$

Volume $\quad V = V^+ \cap V^-$ and the outer boundary is represented by $\partial V = \partial V^+ \cap \partial V^-$

$$\frac{d}{dt}\int_{V\pm} \Phi dV = limit_{h\to 0}\frac{1}{h}\int_{V\pm}\frac{\partial \Phi}{\partial t}dV = \int_{\partial V\pm} \Phi \underline{\dot{x}}\cdot \underline{n}da + \int_S \Phi^+ \mp u_n da$$

$$\frac{d}{dt}\int_{V\pm} \Phi dV = limit_{h\to 0}\frac{1}{h}\int_{V\pm}\frac{\partial \Phi}{\partial t}dV = \int_{\partial V\pm} \Phi \underline{\dot{x}}\cdot \underline{n}da \pm \int_S [\![\Phi]\!] u_n da$$

More generally, with flux term $\mathfrak{J}_\Phi$ and supply term $\sigma_\Phi$ of $\Phi$ the balance and transport equation becomes,

$$\frac{d}{dt}\int_{V\pm} \Phi dV = limit_{h\to 0}\frac{1}{h}\int_{V\pm}\frac{\partial \Phi}{\partial t}dV = \int_{\partial V\pm} \Phi \underline{\dot{x}}\cdot \underline{n}da \pm \int_S [\![\Phi]\!] u_n da + \int_{\partial V} \mathfrak{J}_\Phi \underline{n}da + \int_V \sigma_\Phi dV$$

### Appendix J: Filtering Data Approach for a Power Function

The fitting of data to a function as a polynomial was addressed by Lagrange. Let $P(x) = \sum_{i=1}^{n+1} a_i x^{i-1}$, . At data points $(x_i, f(x_i))$, $P(x_i) = f(x_i) = \sum_{j=1}^{n+1} a_{ij}(x_i)^{j-1}$ . Solving for the coefficients in the expansion is a matrix equation introducing the Vandermonde matrix,

$$\begin{pmatrix} 1 & x_0 & (x_0)^2 & \dots & (x_0)^n \\ 1 & x_1 & (x_1)^2 & \dots & (x_1)^n \\ 1 & x_i & (x_i)^2 & \dots & (x_i)^n \end{pmatrix}\begin{pmatrix} a_1 \\ a_2 \\ a_i \end{pmatrix} = \begin{pmatrix} f(x_0) \\ f(x_1) \\ f(x_i) \end{pmatrix}$$

Alternatively, considering the points as roots of a polynomial equation, a representation may be derived by assuming, $Q(x) = k \cdot \prod_{j=0}^n (x - r_j) = k \cdot (x - r_1)(x - r_2)\dots(x - r_n)$

Form $L_i(x)$  $L_i(x_i)=1$, but for $\neq i$, $L_i(x_j) = 0$, j = 0,..n.     $L_i(x) = \frac{(x-x_0)(x-x_1)...(x-x_n)}{(x_i-x_0)(x_i-x_1)...(x_i-x_n)}$

or in compact notation,    $L_i(x) = \prod_{j=0, j\neq i}^{n} \frac{(x-x_j)}{(x_i-x_j)}$

Now consider a power function representation of kinetics data as a time-dependent function, or a signal in the filter problem. Here fractional differentiation is used to find the function's response from using the system equation.

$$d^\alpha f(x) = \sum_{p=0}^{\infty} a_p \, d^\alpha(x^p) = \sum_{p=0}^{\infty} a_p \frac{\Gamma(p+1)}{\Gamma(p-\alpha+1)} x^{p-\alpha}$$

$$H(z) = \sum_{k=0}^{N} h(k) \, z^{-k}$$

For data or signal f(n) passing through the system the output is given by y(n) such that,

$y(n) = \sum_{k=0}^{N} h(k) f(n-k)$  In the case of a power series such as by a Taylor expansion of the function or linear regression fit,    $f(n-k) = \sum_{p=0}^{N} a_p (n-k)^p$

Then    $y(n) = \sum_{p=0}^{\infty} a_p \sum_{k=0}^{N} h(k)(n-k)^p$   with filter coefficients h(k).

$$y(n) = d^\alpha f(n-\tau)$$

$\sum_{p=0}^{N} h(k)(n-k)^p = \frac{\Gamma(p+1)}{\Gamma(p-\alpha+1)} (n-\tau)^{p-\alpha}$  Using matrix vector notation a linear system of algebraic equations are solved to obtain the impulse response function h.

$A\underline{h} = \underline{b}$ ,    $\underline{h} = A^{-1}\underline{b}$

$$\underline{h} = (h(0), h(1), h(2) ... h(N))^T \qquad A = \begin{pmatrix} 1 & 1.. & 1 \\ N+1 & N.. & 1 \\ (N+2)^2 & N^2 ... & 1 \end{pmatrix}$$

$$\underline{b} = \left(\left(\frac{\Gamma(1)}{\Gamma(1-\alpha)}\right)(N+1-\tau)^{-\alpha}, \left(\frac{\Gamma(2)}{\Gamma(2-\alpha)}\right)(N+1-\tau)^{1-\alpha} ... \left(\frac{\Gamma(N+1)}{\Gamma(N+1-\alpha)}\right)(N+1-\tau)^{N-\alpha}\right)$$

The $A$ matrix is a Vandermonde matrix enabling a calculable inverse for finding the impulse response.

Reference: C. Tseng, IEEE Signal Processing, 8, 3, March 2001, 77.

# Appendix K: Matrix Population Model

At the end of the first time period of a year the population count may be calculated as a multiplying distribution accounting for births and removals of the start or previous year as follows,

$\underline{n}_1 = \underline{\underline{R}}\,\underline{n}_0(t_1 - t_0)$, then $\underline{n}_2 = \underline{\underline{R}}\,\underline{n}_1(t_2 - t_1) = \underline{\underline{R}}^2 \Delta t^2 \underline{n}_0$, and for the year k, $(t_2 - t_1) = (t_1 - t_0) \equiv \Delta t = t_k - t_{k-1}$

$\underline{n}_k = \underline{\underline{R}}\,\underline{n}_{k-1}\Delta t = \underline{\underline{R}}^k \Delta t^k \underline{n}_0$. Summing over the duration of k-years obtains,

$\underline{n}_k = \underline{n}_0 + \underline{n}_1 + \cdots \underline{n}_{k-1}$ Using the distributive property and definition of the exponential of the matrix,

$\underline{n}_k = \underline{n}_0 \left(\underline{\underline{I}} + \underline{\underline{R}}\Delta t + \underline{\underline{R}}^2 \Delta t^2 \ldots \underline{\underline{R}}^k \Delta t^k\right) \quad \underline{n}_k = \underline{n}_0\, e^{\underline{\underline{R}}\Delta t}$

A matrix approach accounting for births, removals by death or migration, and age may be constructed as follows,

$$\begin{pmatrix} r_{F,0} + \beta_{F,0} & 0 & 0 & & \\ 0 & r_{F,1} + \beta_{F,1} & 0 & \cdots & \\ 0 & 0 & r_{F,2} + \beta_{F,2} & & \\ & & \vdots & \ddots & \vdots \\ 0 & 0 & 0 & \cdots & r_{F,k-1}+\beta_{F,k-1} \end{pmatrix} \begin{pmatrix} n_{F,0} \\ n_{F,1} \\ \vdots \\ n_{F,k-1} \end{pmatrix} = \begin{pmatrix} n_{F,1} \\ n_{F,2} \\ \vdots \\ n_{F,k} \end{pmatrix}$$

$$\begin{pmatrix} r_{M,0} + \beta_{M,0} & 0 & 0 & & \\ 0 & r_{M,1} + \beta_{M,1} & 0 & \cdots & \\ 0 & 0 & r_{M,2} + \beta_{M,2} & & \\ & & \vdots & \ddots & \vdots \\ 0 & 0 & 0 & \cdots & r_{M,k-1}+\beta_{M,k-1} \end{pmatrix} \begin{pmatrix} n_{M,0} \\ n_{M,1} \\ \vdots \\ n_{M,k-1} \end{pmatrix} = \begin{pmatrix} n_{M,1} \\ n_{M,2} \\ \vdots \\ n_{M,k} \end{pmatrix}$$

Male and female populations are calculated from birth $\beta$ or removal $r$ by death or other causes information. The fraction of the births $b_i$ that are male and female are respectively, $f_{M,i}$ and $(1 - f_{F,i})$. Source terms may also be introduced as relevant.

$\beta_{M,i} = f_{M,i} b_i \quad b_i = b_{M,i} + b_{F,i}$

$$\beta_{F,i} = (1 - f_{F,i}) b_i$$

For a total population and continuous time model,

$$\underline{\dot{P}} = \underline{\underline{R}}\,\underline{P} + \underline{S}, \qquad P(t) = P_0 \text{ at } t = 0$$

$$\frac{d}{dt}\left[e^{-\int_0^t \underline{\underline{R}}(s)ds}\,\underline{P}\right] = e^{-\int_0^t \underline{\underline{AR}}(s)ds}\,\underline{S}$$

$$\left[e^{-\int_0^t \underline{\underline{R}}(s)ds}\,\underline{P}\right] = \int_0^t e^{-\int_0^t \underline{\underline{R}}(s)ds}\,\underline{S}\,dt + P_0$$

$$\underline{P} = e^{\int_0^t \underline{\underline{R}}(s)ds}\left[\int_0^t e^{-\int_0^t \underline{\underline{R}}(s)ds}\underline{S}(t)\,dt + \underline{P}_0\right]$$

For a constant matrix, $\underline{\underline{R}}(s) = \underline{\underline{R}}$ $\quad \underline{P} = \left[\int_0^t \underline{S}(t)\,dt + \underline{P}_0 e^{\underline{\underline{R}}t}\right]$ and for $\underline{S}(t) = \underline{0}$,

$$\underline{P} = e^{\underline{\underline{R}}t}$$

## Appendix L: A Finite Element Approach for a Fractional Advection Dispersion Model in One Dimension

Using a Caputo differential definition for a fractional operator, the following has been reported in the literature for solving,

$$\frac{\partial C}{\partial t} + v\frac{\partial C}{\partial x} = D\left(\left(\frac{1+\beta}{2}\right)\frac{\partial^{\alpha-1}C}{\partial x^{\alpha-1}} + \left(\frac{1-\beta}{2}\right)\frac{\partial^{\alpha-1}C}{\partial(-x)^{\alpha-1}}\right)$$

$$-1 \leq \beta \leq 1$$

Simple basis functions defined on the intervals of the x axis and linear approximations with integration of parts were used in the derivation, as well as calculus

$$\frac{\tau}{\Delta}GM^+\left(W_0^+ C_0^{k+1} + \sum_{i=1}^{j-2} W_i^+ C_i^{k+1}\right) + \left(-\frac{\tau v}{2\Delta} + \frac{\tau}{\Delta}WA\right)C_{j-1}^{k+1} + \left(1 + \frac{\tau}{\Delta}WB\right)C_j^{k+1}$$

$$\left(\frac{\tau v}{2\Delta} + \frac{\tau}{\Delta}WE\right)C_{j+1}^{k+1} + \frac{\tau}{\Delta}GM^-\left(\sum_{i=j+2}^{N-2} W_i^- C_i^{k+1} + W_N^- C_N^{k+1}\right) = C_j^k \quad j = 1,2,\ldots N-1$$

$$W_i^+ = -(j-i-2)^{3-\alpha} + 4(j-i-1)^{3-\alpha} - 6(j-i)^{3-\alpha} + 4(j-i+1)^{3-\alpha} - (j-i+2)^{3-\alpha}$$

$$W_i^- = -(i-j-2)^{3-\alpha} + 4(i-j-1)^{3-\alpha} - 6(i-j)^{3-\alpha} + 4(i-j+1)^{3-\alpha} - (i-j+2)^{3-\alpha}$$

$$WA = GM^+(-6 + 4(2)^{3-\alpha} - (3)^{3-\alpha}) - GM^-$$

$$WB = GM^+(4 + (2)^{3-\alpha}) - GM^-(4 - (2)^{3-\alpha})$$

$$WE = -GM^+ + GM^-(-6 + 4(2)^{3-\alpha})$$

$$W_0^+ = -(j-2)^{3-\alpha} + 3(j-1)^{3-\alpha} - 3(j)^{3-\alpha} + 4(j+1)^{3-\alpha}, j \geq 2$$

$$W_N^- = -(N-j-2)^{3-\alpha} + 43(N-j-1)^{3-\alpha} - 3(N-j)^{3-\alpha} + (N-j+1)^{3-\alpha} - (i-j+2)^{3-\alpha}$$

$$GM^- = \frac{D(1-\beta)}{2\Gamma(4-\alpha)}\Delta^{1-\alpha}$$

$$GM^+ = \frac{D(1+\beta)}{2\Gamma(4-\alpha)} \Delta^{1-\alpha}$$

Ref. Huang, Quanzhong, Guanhua Huang, and Hongbin Zhan, A Finite element solution for the fractional advection-dispersion equation, Adv Water Res, 31 (2008) 1578-1589.

## Appendix M: Point in Polygon

Large population data sets are generated such as the Census with population count by enumeration district recorded with their centroids. The population data for a defined site area, or around a highway network may be formulated as a point in a polygon for sorting and sifting the population data sets. Subtleties exist in the computer implementation. The concept of this approach depicted below.

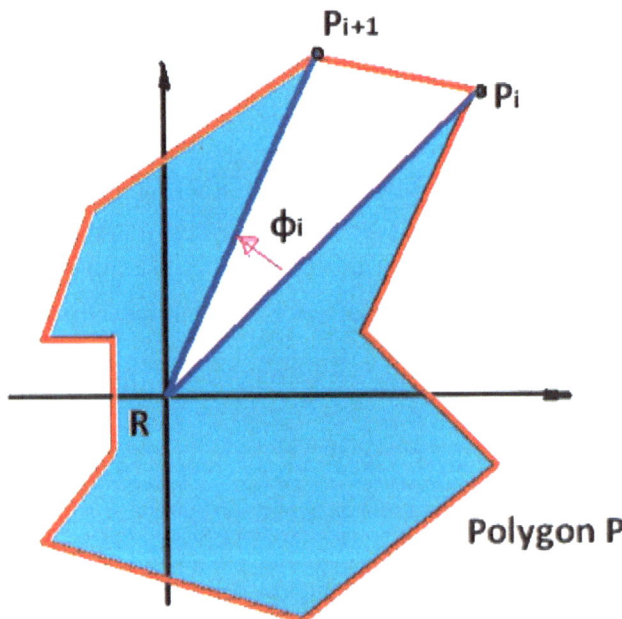

$$\omega(R,C) = \frac{1}{2\pi} \int_a^b d\varphi(t) = \frac{1}{2\pi} \int_a^b \frac{\dot{y}(t)x(t) - y(t)\dot{x}(t)}{x(t)^2 + y(t)^2} dt$$

Consider the array of n points, $P_0, P_1, \ldots P_{n-1}, P_n = P_0$ and a piecewise linear curve,

$(x_i(t-i), y_i(t-i))^T \quad t \in [i, i+1] \quad (x_i(t), y_i(t))^T = tP_{i+1} + (1-t)P_i$

$R = (0,0)^T, \; P = (P_x, P_y)^T, \; Q = (Q_x, Q_y)^T, \; \cos\alpha = \frac{\langle P|Q\rangle}{\|P\|\|Q\|}, \; \cot\beta = \frac{\langle P-Q|P\rangle}{|D|}, \; \cot\gamma = \frac{\langle Q-P|Q\rangle}{|D|},$

$$D = \det\begin{pmatrix} P_x & P_y \\ Q_x & Q_y \end{pmatrix} \begin{pmatrix} x(t) \\ y(t) \end{pmatrix} = tQ + (1-t)P, \ t \in [0,1]$$

$$\omega(R,C) = \int_0^1 \frac{D}{t^2\langle Q-P|Q-P\rangle + 2t\langle Q-P|P\rangle + \langle P|P\rangle}$$

$$= \arctan\frac{\langle Q-P|Q\rangle}{D} + \arctan\frac{\langle P-Q|P\rangle}{D} = sign(D)(arctancot\gamma + arctancot\beta)$$

$$sign(D)(\pi - \gamma - \beta) = sign(D)\alpha$$

$$\omega(R,C) = \frac{1}{2\pi}\sum_{i=0}^{n-1} A_i \cdot \arccos\frac{\langle P_i|P_{i+1}\rangle}{\|P_i\|\|P_{i+1}\|} \text{ where } A_i = sign\begin{vmatrix} P_i^x & P_{i+1}^x \\ P_i^y & P_{i+1}^y \end{vmatrix}$$

$$\omega(R,C) = \frac{1}{2\pi}\sum_{i=0}^{n-1}\varphi_i$$

$\varphi_i$ is the signed angle between $\overline{RP_i}$ and $\overline{RP_{i+1}}$ This computes a winding number of a point with respect to a polygon as the number of revolutions while traversing along P once and R is not intersected. Derivation and discussion of computational approaches to make an efficient algorithm for spatial location analysis. The reference compares the similarity and differences with an alternative ray crossing even-odd approach.

Ref: Hormann, Kai and A. Agathos, The point in polygon problem for arbitrary polygons, Computational Geometry 20 (2001) 131-144.

# Part II.

System Reliability of Nuclear Power Generation: Site Risk and Engineering Design

ABSTRACT:

The aim of this reporting is to present quantitative perspectives on system reliability including alternative comparisons of the safety in view of both site and probable engineering design vulnerabilities. Here the radiological risks to the surrounding population for existing sites located in two major regions of the United States, the mid-Atlantic and Northeastern regions are investigated for new generation and possible design improvement. Statistical z-scores, clustering methods by both k-means, and agglomeration algorithms, as well as by practical mathematical optimization implementations are applied. Sites are compared with assorted population, exposure, and risk indices to compare sites as part of the overall system comprising site and technology. Additionally, a time-dependent discrete probability model is simply derived with present value calculations for comparing cost allocations for improving system reliability or for evaluating aging and extending service lifetime. These comprise a class of probabilistic rare event jump processes which are significant in achieving system safety shown here with fractional calculus.

INTRODUCTION:

The safety of nuclear steam electric plants resides in many areas, including the site, design, operation, and emergency preparations for probable severe consequence incidents. The purpose of this paper is to focus on site aspects, or location analysis of new additional generating capacity [1,2], as well as address financial analysis of existing design change options. Probabilistic reliability engineering tools [44], a safety culture [38]., and decisions on resource allocations for land use as part of the system for providing energy is supported by simulation models. The nuclear energy option with certainty offers the benefits of obviating greenhouse gas warming over fossil fuels. Realistic energy plans with land use will continue to rely on both, along with alternative technologies. The analysis utilizes safety goals [5,17,19], and formulates reliability improvement costs [7] or cost surrogate measures of rare severe consequence accidents.

In the United States, primarily as a result of Three Mile Island accident (Unit 2 on March 28, 1979) significant changes have been implemented to improve emergency preparedness, including planning efforts, training of people among organizations, and development of technology aids. The use of probabilistic risk assessment for regulatory guidance, for the development of safety goals, and for extensive research on severe accident risks to offsite populations have extended the range of safety considerations from *normal* to a broader class of accident conditions [6, 27]. Probabilistic Risk Assessment (PRA) complements the traditional defense-in-depth philosophy to reduce unnecessary conservatism associated with current regulatory requirements and guides, license commitments, and back-fit evaluations.

Even-though studies confirm no increased risk from cancer to surrounding populations in 107 counties adjacent to 62 nuclear facilities from normal operation, uncertainties in safety remains. The health impacts from the Three Mile Island fuel melting accident emitting releases spread by wind around 360 degrees of the compass can only be estimated. It has been calculated that the maximum cumulative off-site radiation dose to an individual was less than 100 mrem ( mSv). The total population exposure has been estimated to be in the range from about 1000 to 5000 person-rem. This exposure could statistically produce zero or one additional fatal cancer over the lifetime of the exposed population of approximately 2 million surrounding the site.

On April 26, 1986, a major accident occurred at reactor 4 of the Chernobyl Nuclear Power Station in the Soviet Union releasing radioactivity which deposited throughout the Northern Hemisphere. It is estimated that approximately, 800 million people account for 97 percent of the total risk increment. Beyond the 19-mile (30-km) zone surrounding Chernobyl, the incremental increase in cancer fatality risk is estimated

to be a fraction of a percent. It is noted that the Chernobyl accident's radioactivity release was comparably the same in magnitude, for the types of radioactivity released as a US nuclear plant design, validating earlier research and models. Severe accidents for light water reactors of United States' design, however differ from the Chernobyl reactor

in an important way. The light water reactor design has a negative void reactivity response that is opposite to that of a Chernobyl design where voids from boiling would substantially increase the reactivity and course to damaging fuel. The role of a containment in a severe accident gained focus during both Three Mile Island and Chernobyl accidents. Many countries have responded with supplemental containment volume, filtering, and venting strategies.

. In general, information can be categorized as nominal, ordinal, or cardinal. That is, nominal information represents a descriptive approach, while ordinal information represents ranks, and cardinal data are numerical, usually with real numbers. A mix of all three kinds of information comprises documentation accompanying nuclear site environmental impact and safety review reports. Many useful insights have been gained from ranking of risks of multiple hazards to small communities [23], in addition to operating plant experience [43]. The USNRC's Safety Goals with quantitative objectives are to be used with consideration of uncertainties in making regulatory judgments on nuclear power plant licensees. The policy is intended to allow the many applications of PRA to regulatory decisions, making the regulatory process risk informed for the use of risk insights to direct focus on protection of public health and safety to possibly reduce residual risks. This regulatory history has motivated these investigations on both siting and cost considerations for finding alternatives for siting and for making improvement to existing site facilities.

Radiological consequences of severe radiological accidents releasing radioactivity to the atmosphere are assessed by pathways of cloud immersion or shine, by ground deposited radioactivity shine, or by inhalation either directly, or from a resuspension of radioactive particulates. The consequences for these most severe accidents have radiological impacts in regions far from the site [18, 20, 25,29,36].

Furthermore, the first accident models pertained primarily to the early phase of an emergency followed by additional modeling and field data to address the dispersal of the contamination and additional pathways such as ingestion producing longer effects. Many of the pathways for radiation and the transport of radioactive materials that lead to radiation exposure hazards to humans are the same for severe accidental as for "normal" releases.

1.1 Planning Regions: Mid-Atlantic and Northeastern United States

This analysis addresses population within several radial distances, without or with wind direction frequency weighting as an exposure index which is illustrated with these selected nuclear site examples. The nuclear power plant electricity generation sites analyzed in the northeastern United States are listed in the table below.

|    | Nuclear Site |
|----|--------------|
| 1  | Beaver Valley 1,2 |
| 2  | Limerick 1,2 |
| 3  | Peach Bottom 1,2 |
| 4  | Susquehanna 1,2 |
| 5  | Three Mile Island 1 |
| 6  | Calvert Cliffs 1, 2 |
| 7  | Hope Creek |
| 8  | Oyster Creek |
| 9  | Pilgrim |
| 10 | Vermont Yankee |
| 11 | Millstone |
| 12 | Seabrook |
| 13 | Fitzpatrick |
| 14 | Ginna |
| '15 | Indian Pt |
| 16 | Nine Mile |
| 17 | Shoreham |
| 18 | Davis Besse |
| 19 | Perry |
| 20 | Maine Yankee |

Table 1: Listing of Nuclear Sites in Planning Region:

The total population within each fifty mile circular radii of the sites is charted. *The m*iddle year of license renewal period rounded up to the next year for which population forecasts were used.
Probabilistic risk and several siting criteria addressing population are described next, followed by calculated results for population and risk measures, cost estimations, and concluding remarks..

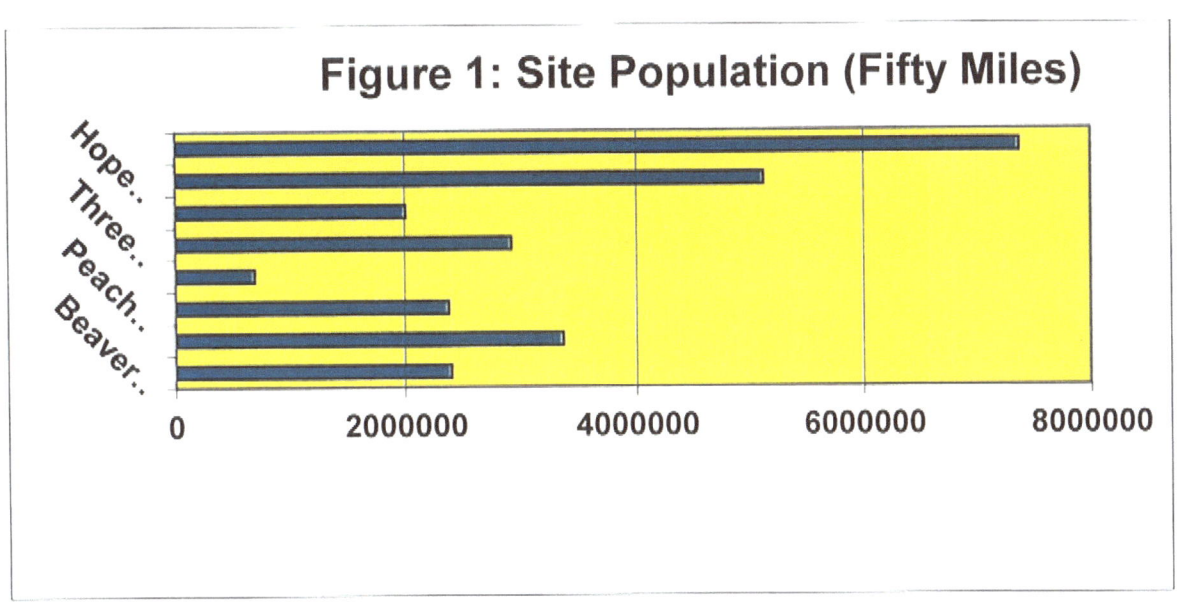

Figure 1: Site Populations ( 50 mi = 80 km)

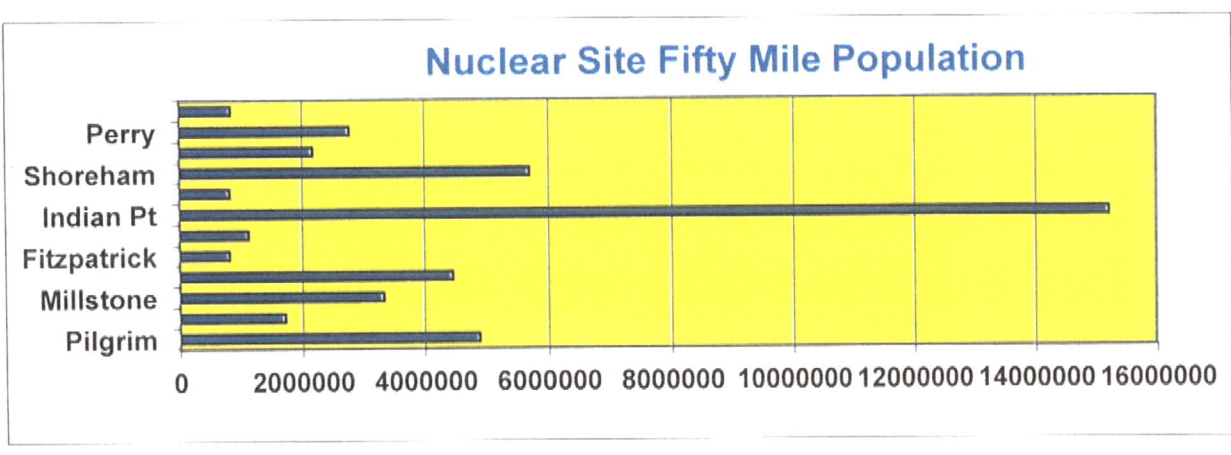

Figure 2: Site Populations ( 50 mi = 80 km)

The US Nuclear Regulatory Commission's (USNRC) Probabilistic Risk Policy addresses use of probabilistic assessment in its regulatory matters to complements the traditional defense-in-depth approach. The intent is to reduce unnecessary conservatism associated with current regulatory requirements and guides, license commitments, and staff practices. PRA (Probabilistic Risk Assessment) should be used to support proposals for additional regulatory requirements where appropriate in accordance with the USNRC's backfit rule, 10 CFR 50.109, which describes the process necessary to support imposition of additional regulatory requirements. PRA evaluations in support of regulatory decisions should be as realistic as practicable, and appropriate supporting data should be publicly available for review. The commission's safety goals and subsidiary numerical objectives are to be used with consideration of uncertainties in making regulatory judgments on the need for proposing and back-fitting new generic requirements on nuclear power plant licensees. The commission's policy is intended to allow the many applications of PRA to be implemented in a consistent and predictable manner that would promote regulatory stability, efficiency, and predictability of regulatory decisions, making the regulatory process risk informed (the use of risk insights to focus on those items most important to protecting public health and safety).

## 1.2 Site Criteria

Siting criteria for the United States enumerated in the Code of Federal Regulations, 10 CFR Part 100—Reactor Site Criteria (US NRC) are summarized below for determination of an exclusion area, a low population zone, and identification of population centers' distances. Each reactor site must have certain characteristics that tend to reduce the risk and potential impact of accidents.

*Exclusion area* means the area surrounding the reactor in which thereactor licensee has the authority to determine all activities includingexclusion or removal of personnel and property from the area. This area may be traversed by a highway, railroad, or waterway, providedthese are not so close to the facility as to interfere with normaloperations of the facility, and provided appropriate and effectivearrangements are made to control traffic on the highway, railroad,

or waterway, in case of emergency, to protect the public healthand safety. Residence within the exclusion area shall normally beprohibited. In any event, residents shall be subject to ready removal in case of necessity. Activities unrelated to operation of the reactor may be permitted in an exclusion area under appropriate limitations, provided that no significant hazards to the public health and safety will result.

*Population center distance* means the distance from the reactor to the nearest boundary of a densely populated center containing more than about 25,000 residents.

*Low population zone* means the area immediately surrounding the exclusion area which contains residents, the total number and density of which are such that there is a reasonable probability that appropriate protective measures could be taken in their behalf in the event of a serious accident. These guides do not specify a permissible population density or total population within this zone because the situation may vary from case to case. Whether a specific number of people can, for example, be evacuated from a specific area, or instructed to take shelter on a timely basis, will depend on many factors such as location, number and size of highways, scope and extent of advance planning, and actual distribution of residents within the area.

Determination of exclusion area, low population zone, andpopulation center distance.

High population density sites offer the reduced costs of transmission to meet the demand for electricity, however conflicting safety risk considerations driven by population would lead to remote areas for nuclear sites. Selected studies using screening techniques, or including mathematical programming for regional planning, or addressing general nuclear siting issues are reported.

## 2 SITE POPULATION AND RISK MEASURES

Accident consequence modeling has provided both quantitative and qualitative insights for constructing indices for site safety evaluation. The radiation exposure (hazard) to individuals is determined by the individual's proximity to the radioactive materials A site population factor method for weighting population radially in a circular polar grid has been a traditional practice for screening sites. The weighting is an exponential function of range to the negative 1.5 power to generally account for atmospheric dilution for long-range transport.

## 2.1 Population Sector Guides

The United Kingdom's (UK) Health and Safety Executive Nuclear Directorate apply land use planning limits both sectorially and cumulatively to an outer radius R from the site [40]. These limits are here derived for United State's site population exposure criteria at start-up and throughout plant lifetime using sixteen sector (22.5°) compass rather than the UK's twelve.

$$\int_{r=1}^{R} \frac{D}{r^{1.5}} \frac{2\pi \cdot r \cdot dr}{16} = SL \quad \text{or} \quad SL = \frac{\pi D}{4}\left(R^{.5} - 1\right) \quad (1)$$

Cumulative Site Limit (Outer radius R):

$$\int_{r=1}^{R} \frac{2\pi D \cdot r \cdot dr}{r^{1.5}} = CSL \quad \text{or} \quad SL = 4\pi D\left(R^{.5} - 1\right) \quad (2)$$

Table I: Sector Population Exposure Guides Eq(1)

| Pop density / mi^2 | Radius (miles)→10 | 50 | 150 |
|---|---|---|---|
| 500 | 849.12 | 2384.10 | 4416.86 |
| 1000 | 1698.25 | 4768.20 | 8833.72 |

Table II: Cumulative Population Limits Eq(2)

| Pop density / mi^2 | Radius (miles)→10 | 50 | 150 |
|---|---|---|---|
| 500 | 13585.98 | 38145.61 | 70669.74 |
| 1000 | 27171.959 | 76291.22 | 141339.5 |

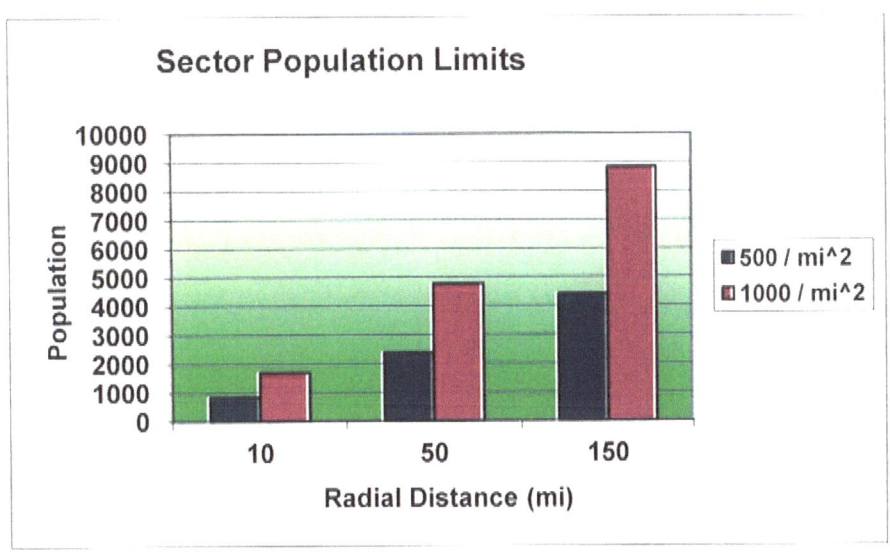

Figure 3: Sector Population Exposure (1 mi = 1.609 km, 1 mi^2 = 2.589 km^2)

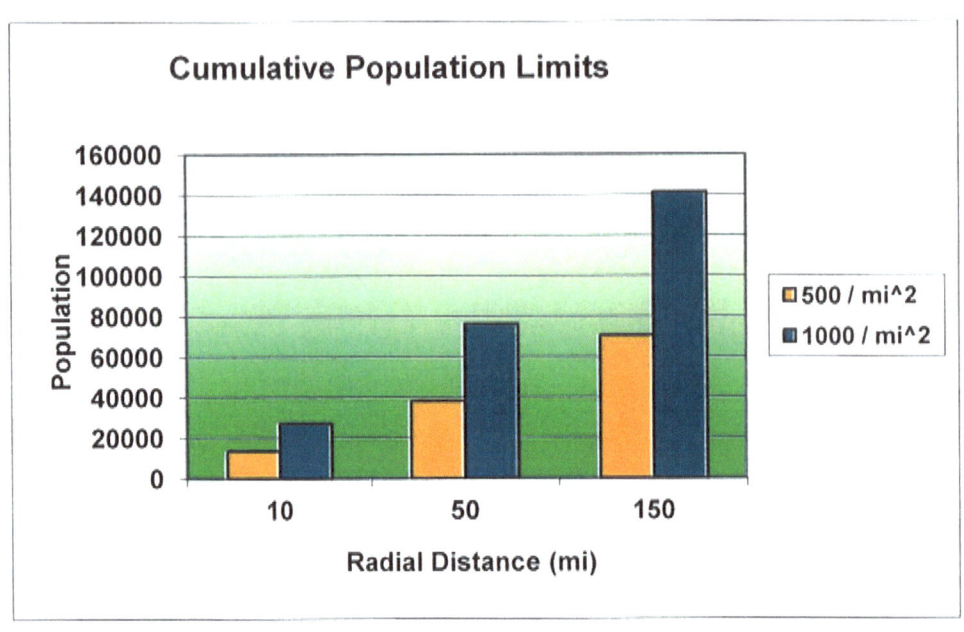

Figure 4: Cumulative Population Limits: (1 mi = 1.609 km, 1 mi^2 = 2.589 km^2)

The 500 and 1000 people per square-mile population densities to several radii are shown in Tables II and Table III for Sector and Population Exposure Guides. This can easily be modified for several compass sectors to accommodate wind persistence. Tables I and II represent the total population for the defined areas assuming a uniform population density D.

2.2 Site Risk Indices

The following site indices emphasize different site characteristics for use in comparing sites for which illustrative calculations are given.

2.3 Statistical z-Score

Well known is the fact that the expected value or mean value of a data set captures a distribution's center-of-balance for a single variable function. Statistical z-scores were theoretically computed which indicate how far and in what direction, that risk measure deviates from its distribution's mean, expressed in units of its distribution's standard deviation [21].

The z-Score is defined for data i by,

$$z = \frac{x_i - \mu}{\sigma} \qquad (3)$$

2.4 Exposure Index EI

Siting perspectives from severe accidents and technical guidance with recent generic license renewal studies have motivated development of Exposure Indices which weight population by the annual frequency of wind direction occurrence in each compass sector [44]. Both the calculation for ranges of R= 10 miles and R= 150 miles are used.

$$EI = \sum_{j=1}^{16} \sum_{i=1}^{R} w_j Pop(i,j) \qquad (4)$$

## 2.5 Reactor Power * EI

This is calculated by multiplication with the reactor power generation captured with net generation data. The intent is to construct a measure that scales with release inventory, in particular gamma emitting radionuclides as a surrogate measure for source term magnitude, which along with population dominate offsite consequences,

$$EI = \sum_{j=1}^{16} \sum_{i=1}^{R} w_j Pop(i,j) * NetGen \qquad (5)$$

## 2.6 Site Safety Goal

The USNRC Safety Goal for latent cancer numerically estimates the risk to an average individual within fifty miles from the site which shall not exceed 0.1 % of the national latent cancer risk. This quantifies to approximately $1 \cdot 10^{-6}$ using vital statistics. The fifty mile population was multiplied by this factor and then divided by the product of the calculated population limit and this factor. This obtains a unit-less index that essentially is the same as the quotient of the total site population and the limiting population based on $1 \cdot 10^{3}$ persons per square mile criteria. In general, impacts reach beyond 100 miles radially for the most severe accidents, as well as are modified by the time-series of site meteorological conditions and resulting effects on range-dependent consequences. The values for total dose calculated by MACCS and CRAC2S are robust in comparison, differing by less than a factor of two.

## 2.7 Windy Population Sector.

The sector with the highest annual frequency of wind blowing in that direction is chosen to accumulate the population to fifty miles.

## 3.0 Calculated Results

The following displays several of the site risk measures in terms of z-Scores and then followed by clustering analysis.

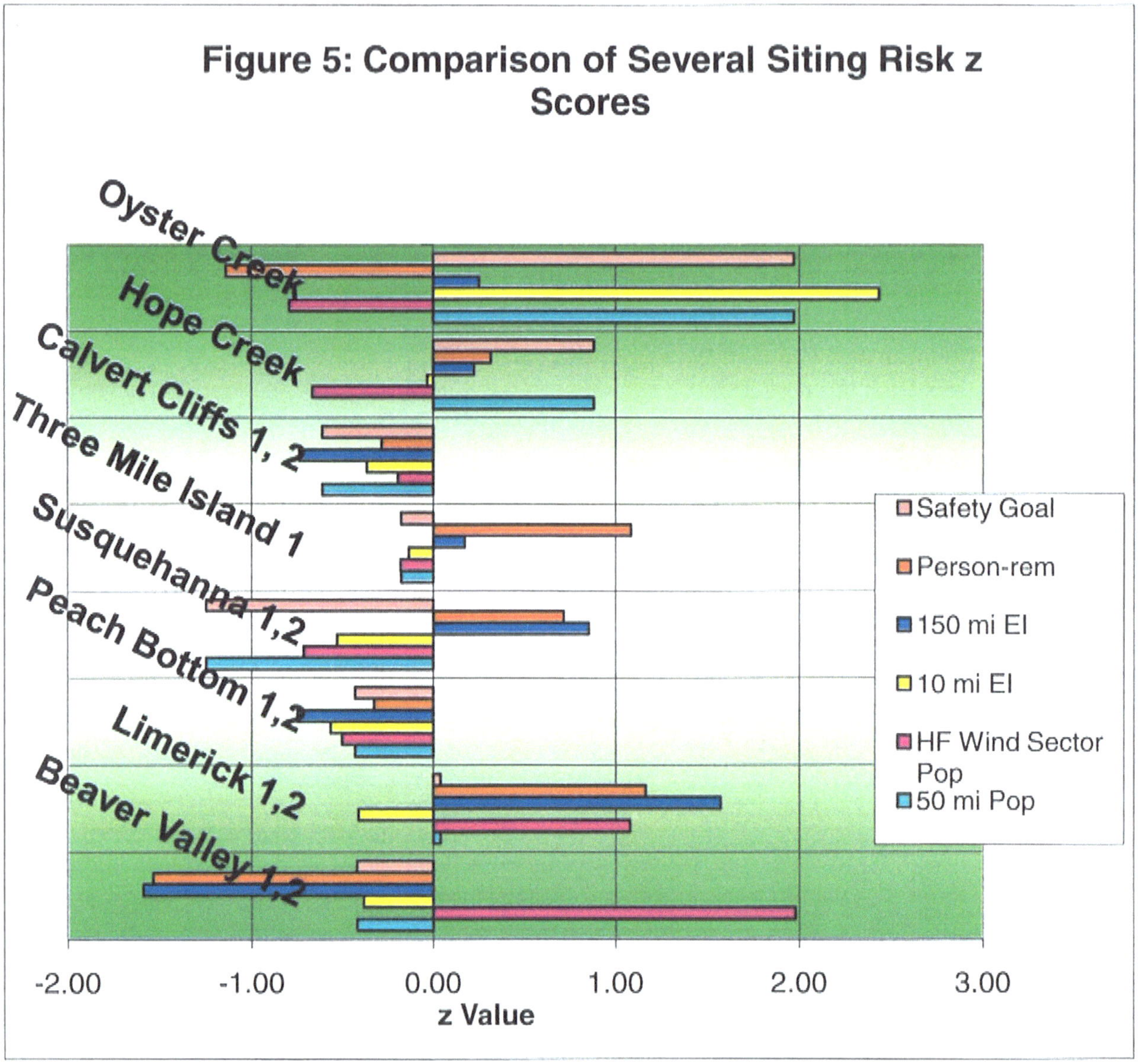

**Figure 5: Comparison of Several Siting Risk z Scores**

Figure 5: Nuclear Power Plant Site Comparisons Using z-Scores

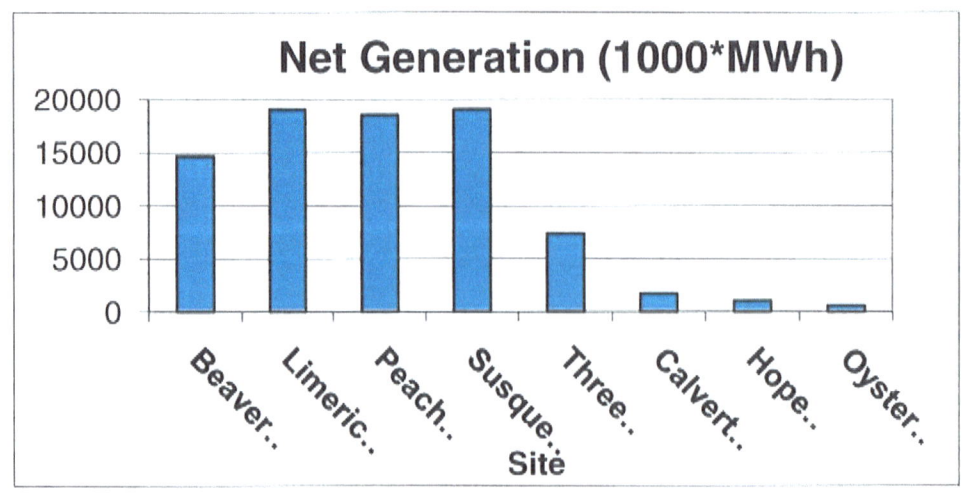

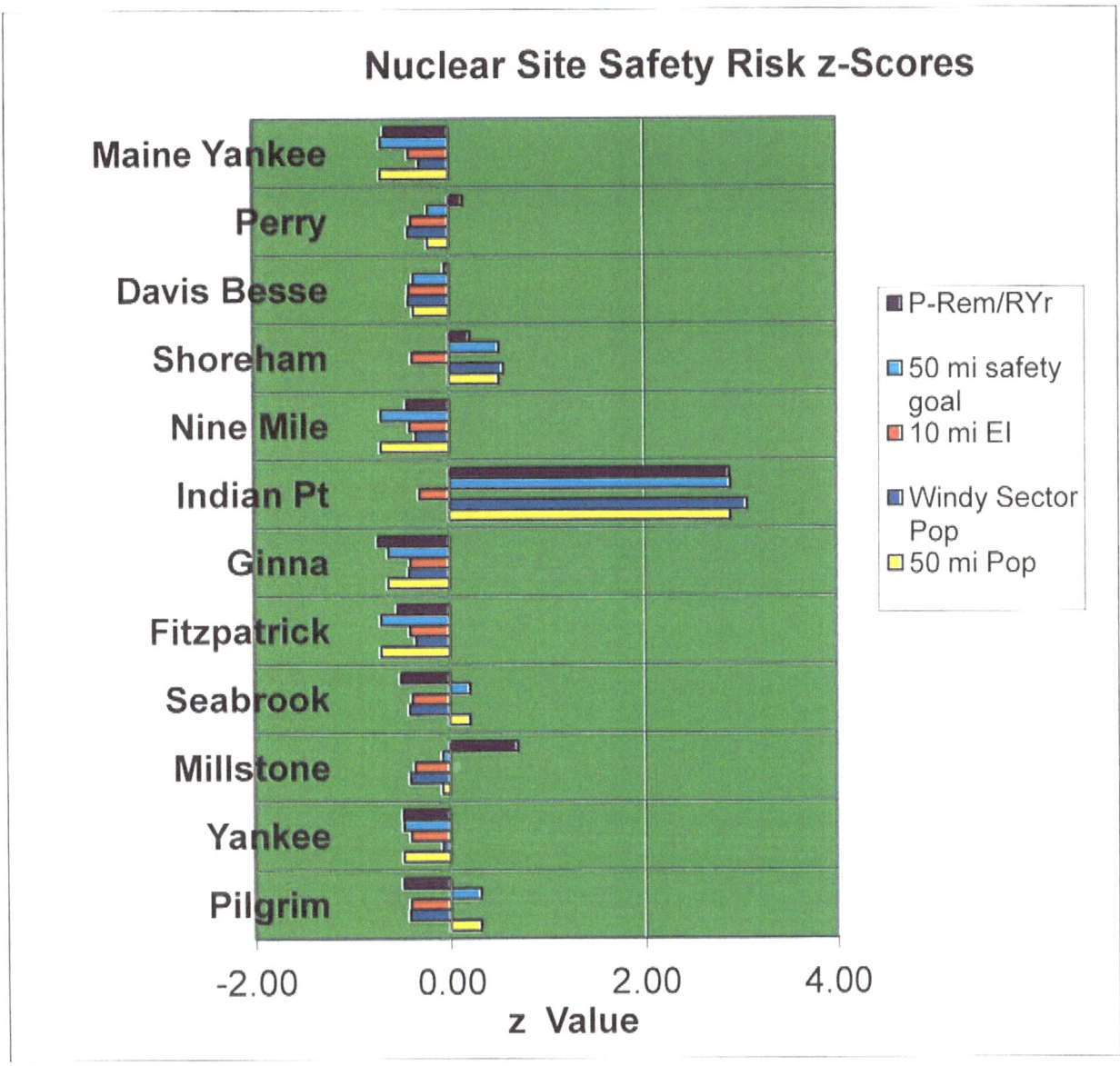

Figure 7: Nuclear Power Plant Site Comparisons Using z-Scores

Samples from a Normal Distribution were used to simulate uncertainties in the Site Indices from which descriptive statistics were derived. The averages, maximums, minimums, and z-scores are displayed in the following figures for the Northeastern regional sites.

Table III: 50 Mile Populations using Clustering: k-Means, Trace(Median)

| Risk Cluster 1 | Risk Cluster 2 | Risk Cluster 3 | Risk Cluster 4 | Risk Cluster 5 |
|---|---|---|---|---|
| Beaver Valley | Limerick | Seabrook | Calvert Cliffs | Hope Creek |
| Susquehanna | Three Mile Island | Peach Bottom | Oyster Creek | Ginna |
| Vermont Yankee | Pilgrim | Fitzpatrick | Nine Mile | |
| Millstone | Shoreham | Indian Point | | |
| | Davis Besse | Perry | | |

Table IV: 50 Mile Populations Using Clustering: Agglomerative

| Beaver Valley | Susquehanna | Hope Creek | Indian Point |
|---|---|---|---|
| Limerick | Fitzpatrick | Oyster Creek | |
| Peach Bottom | Ginna | Pilgrim | |
| Three Mile Island | Nine Mile | Seabrook | |
| Calvert Cliffs | Maine Yankee | Shoreham | |
| Vermont Yankee | | | |
| Millstone | | | |
| Davis Besse | | | |
| Perry | | | |

Agglomerative Hierarchical Clustering is an iterative process that starts by calculating the dissimilarity between the N objects. Then two objects which when clustered together minimize Ward's agglomeration criterion [42], thereby clustering together creating a class comprising these two objects. Then the dissimilarity between this class and the N-2 other objects is calculated iteratively in a similar manner until all the objects have been clustered. Software XLSTAT was used for this purpose.

Table V: 150miEI*NetGen Agglomeration (Ward's criterion)

| Pilgrim | Yankee | Millstone | Fitzpatrick | Indian Point |
|---|---|---|---|---|
| Seabrook | Ginna | | Shoreham | |
| Nine Mile | Davis Besse | | | |
| | Perry | | | |

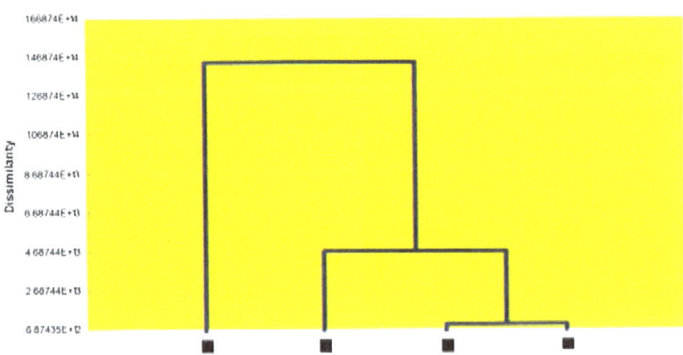

Figure 8: Agglomerative Clustering 50 mi (80 km) Population (Ward's Method) Dendogram

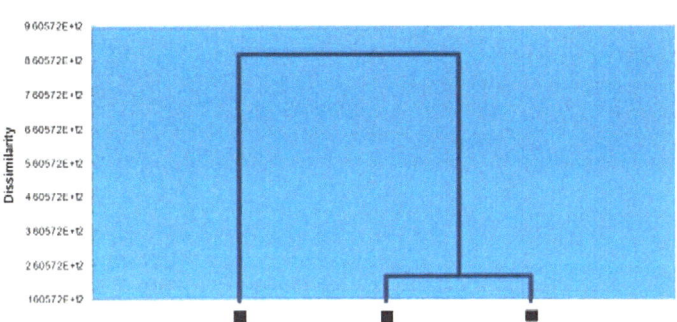

Figure 9: Agglomerative Clustering 150 mi (240 km) EI (Ward's Method) Dendogram

## 3 PROBABILISTIC RISK FORMULATION OF COST

The present value risk calculations using the discrete formulation are displayed and Include compounding annually for a range of expected person-rem valued at $ / person-rem for a term of, say fifteen, twenty, or thirty years. The assumptions of interest rate, compounding interval, and term are for illustration and can easily be modified. The derivation of the calculation formulas with a general probabilistic formulation is summarized below to provide perspectives on costs for existing plant options such as improved design.

The time value of money and risk is computed by

$$PV = \frac{\sum_{d=0}^{D-1} R(d) \cdot \left(1 + \frac{i}{n}\right)^{D-d}}{\left(1 + \frac{i}{n}\right)^{D}} = R(0) + R(1)\left(1 + \frac{i}{n}\right)^{-1}$$

$$+ R(2)\left(1 + \frac{i}{n}\right)^{-2} + R(3)\left(1 + \frac{i}{n}\right)^{-3}$$

$$+ \ldots R(D-1)\left(1 + \frac{i}{n}\right)^{-D-1} \tag{6}$$

where $PV$ = Present value risk, $D$ = Total number deposits, $d$ = deposit index, $i$ = annual interest rate, $n$ = number of times compounded per year, $FV$ = Future value, and $R(d)$ = value of d-th "deposit". This becomes, $R(i) = R \; \forall i = 0, D-1$, The PV formula is a geometric progression where the common ratio r of adjacent terms is defined as $r \equiv \left(1 + \frac{i}{n}\right)^{-1}$,

For a continuous stream of payments, using a limit approach:

49

$$\lim it \Big|_{N\to\infty}\left(1+\frac{1}{N}\right)^N = e \quad \text{where} \quad \frac{1}{N}=\frac{i}{n}; \quad \text{also,} \quad n\cdot(D-d)=i\cdot N\cdot(D-d)$$

so that
$$\lim it \Big|_{N\to\infty}\left(1+\frac{1}{N}\right)^{N\cdot i(D-d)} = e^{i(D-d)}$$

This obtains for the present value risk for T years time interval.

$$PV = \int_{t=0}^{T} R(t)\cdot e^{-i\cdot t} dt \tag{8}$$

Consider during the time interval d (one year) the probability of n radiological releases to the environment given by,

$$P(N,d) = \sum_{j=1}^{K} P_{R_j}(d, D_j) = \sum_{j=1}^{N} P_{R_j}(d|D_j) P_{D_j} \tag{9}$$

$P_{R_j}(d|D_j)$ Probability of Release Conditional on Core Damage State j

$P_{D_j}$ Probability of Core Damage State j

$P_{R_j}(d, D_j)$ Probability of Release during Damage State j

$$P(N=n;d) = \prod_{d=1}^{D} \frac{\left(\int_{d}^{d+1}\rho_\lambda d\rho\right)^{n_d} e^{-\int_{d}^{d+1}\rho_\lambda d\rho}}{n_d!}, \tag{10}$$

$$\left[\frac{\left(\int_{0}^{1}\rho_\lambda d\rho\right)^{n_1} e^{-\int_{0}^{1}\rho_\lambda d\rho}}{n_1!}\right]\cdots\left[\frac{\left(\int_{D-1}^{D}\rho_\lambda d\rho\right)^{n_D} e^{-\int_{D-1}^{D}\rho_\lambda d\rho}}{n_D!}\right]$$

$n = \sum_{d=1}^{D} n_d$. For a constant rate of occurrence, the probability density $\rho_\lambda = \rho$ radiological release events/year, the probability of n events obtains

$$P = \prod_{d=1}^{D}\left[\frac{(\rho_\lambda)^{n_d} e^{-\rho\cdot 1}}{n_d!}\right] = \left(\frac{\rho_\lambda^n e^{-\rho\cdot D}}{(n_D!)!}\right).$$

The probability of at least one event is calculated by

50

$$P(N \geq 1; d = 1, D) = 1 - P(n = 0) = 1 - e^{-\rho \cdot D}$$
$$\approx 1 - (1 - \rho \cdot D) = \rho \cdot D,$$

for $\rho \cdot D \ll 1$. Typical substituted values yield $(10^{-4}) \cdot 40 \ll 1$. In this approximate case the probability is equal to the rate.

Then a simple calculation can be made for

$$R(d) = \rho \cdot CEPR \cdot \frac{V}{person - rem} \qquad (11)$$

where $CEPR$ denotes the Conditional Expected Person-Rem, and $V$ is a valuation such as $ 2000 - $5000 used in regulatory considerations of back-fits to plant design [31].

Radiological dose models for severe accidents calculate this conditional expected person-rem which is typically multiplied by a probability estimate from fault tree and event tree analyses for a radiological release to the environment using this approximation.

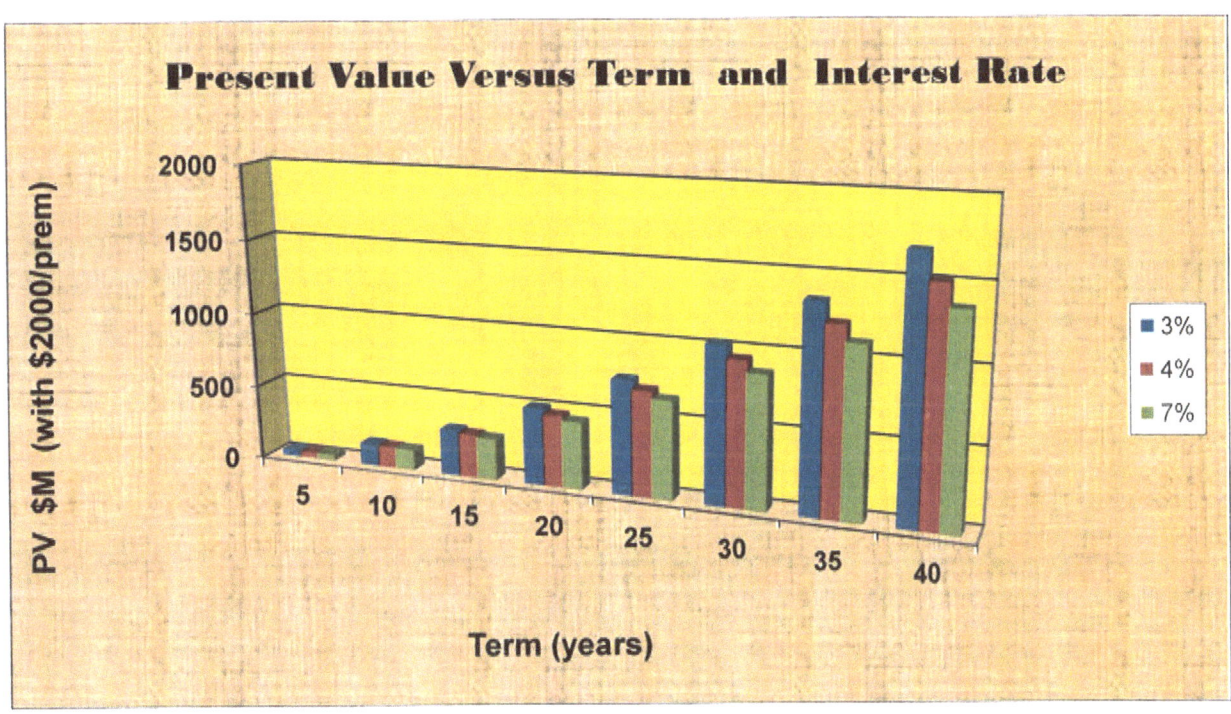

Figure 10: Present Value Risk Costs For Improvement ($2000 / person-rem averted and 1500 expected person-rem averted)

### 3.1 Renewal theory with Fractional calculus

Renewal theory is used to model the counting of events with non-exponential interarrival distributions. Consider a random variable $X_n$ $n = 1, 2, ...$ with common distribution and probability density functions, respectively $F$ and $f$. $X_n$ is the interarrival time between $(n-1)$ and n-th event. Assume $F(0)=0$ $\Pr(X_n=0)<1$, $\mu = E[X_n] > 0$ Define

$$S_n = \sum_{i=1}^{n} X_i,$$

$S_0 = 0$, corresponds to no event occurs at time zero. $N(t)$ represents the number of renewals in time interval $[0,t]$. $N(t) = \max\{n, S_n \leq t\}$. The n-th event occurs at at time $\leq t$ if and only if, the number of events until time t is n.

$$p_n(t) = \Pr[N(t) = n]$$

$$\Pr[N(t) \geq n] - \Pr[N(t) \geq n+1]$$

$$\Pr[S_n \leq t] - \Pr[S_{n+1} \leq t] = F_n(t) - F_{n+1}(t)$$

$$F_n(t) = F_n(t) * F_{n+1}(t) = \int_0^\infty F_n(t-u) f_{n-1}(u) du \quad \text{or an n-fold}$$

convolution of F with itself.

A mean renewal process $M(t) \equiv E[N(t)] = \sum_{n=1}^{\infty} n \Pr[N(t) = n]$ may be examined.

$$\sum_{n=1}^{\infty} n[F_n(t) - F_{n+1}(t)] = \sum_{n=1}^{\infty} F_n(t)$$

This derives $M(t) = F(t) + \int_0^t M(t-x) f(x) dx$ called the renewal equation.

The classical Poisson distribution of rare events may be generalized to a fractional case calculus case. The Poisson describes, for example, the probability of n events in time duration t, by:

$p(n) = \dfrac{(\lambda t)^n e^{-\lambda t}}{n!}$. Now consider $p(k;t) = \dfrac{t^{k\alpha} E_\alpha^{(k)}(-t^\alpha)}{k!}$

for the density function [24] and for an Erlang fractional density by

$$p(k;t) = \alpha \frac{t^{k\alpha} E_\alpha^{(k\alpha-1)}(-t^\alpha)}{(k-1)!}$$

The cumulative distribution function for the latter obtains,

$$P(t) = 1 - \sum_{n}^{k-1} \alpha \frac{t^{na} E_\alpha^{(na-1)}(-t^\alpha)}{(n-1)!}$$

where $E_\alpha(z) \equiv \sum_{n=0}^{\infty} \frac{z^n}{\Gamma(\alpha \cdot n + 1)}$, $\alpha > 0$, and $E_\alpha^{(k)}(z) = \frac{d^k}{dz^k} E_\alpha(z)$

The equation for a continuous time random walker yields,

$$p(x,t) = \delta(x)\Psi(t) + \int_0^t \phi(t-t')\left[\int_{-\infty}^{\infty} w(x-x')p(x',t')dx'\right]dt'$$

The first term represents a persistence (typically decaying in time) of initial position (x=0). The space time convolution in the second term with densities $\phi(t)$ and $w(x)$ gives the probability contribution from $x'$ at time $t' \prec t$ jumping to point $x$ at time $t$, after a wait period from $(t-t')$. The useful case of a compound probability derives for $\phi(t) = e^{-\lambda t}$,

$$\frac{\partial p(x,t)}{\partial t} = -p(x,t) + \int_{-\infty}^{\infty} w(x-x')p(x',t')dx'dt'$$

$$p(x,t) = \sum_{n=0}^{\infty} \frac{(\lambda t)^n e^{-\lambda t}}{n!} w_n$$ . The classical Kolmogorov-Feller equation generalizes to a fractional differential

$$\frac{\partial^\alpha p(x,t)}{\partial t^\alpha} = -p(x,t) + \int_{-\infty}^{\infty} w(x-x')p(x',t')dx'dt'$$

with solution given by

$$p(x,t) = E_\alpha(-t^\alpha)\delta(x) + \sum_{n=1}^{\infty} \frac{t^{\alpha n}}{n!} E_\alpha^{(n)}(-t^\alpha) w_n(x)$$

An optimization of siting, for example $n$ new reactors at the sites in this region of study may be found as a sorting and selection heuristic approach. Using the data the following calculations on each existing sites

j = 1,2..k are made, followed by sorting the calculated data sets for each j, say in ascending values and then selection to achieve the minimum values from the choices presented [4]. This dynamic programming approach is simple and perhaps more closely matches the thinking of Bellman. ANOVA calculations showed a good fit between the full code latent cancer calculations and the exposure indices. Further, a Wilcoxon test between the two regions' sites computed a p-value (0.78) that is greater than the significance level alpha (0.05), indicating one cannot reject the null hypothesis that the site fifty mile populations are the same. A regression of the population and calculated offsite health impacts in rem(.01 Sv) per year from severe accidents obtained an $R^2 = 0.83$. Consider here k = 12 and define

$ES(j)$ = existing site measure such as the average 150-mile Net Generation *Exposure Index simulation is made by selecting random samples from probability distribution model. $NES(j)$ = a standard amount of new electrical generation supply for planning purposes, say 500 MWh Then the optimization is formulated as minimizing

$$\text{Min } \{TS_n(j)\} = ES(j) + n \cdot NES(j) \quad n = 1,..k$$

Solutions for n = 1,2, 3, 4 new reactors generating 500 MWh can easily be listed. After sorting the $TS_1(k)$, $TS_2(k)$, $TS_3(k)$, $TS_4(k)$ into a ranking order k = 1 to k, here k = 1 being the smallest value, k+1 the next smallest in the list, through the kth value (maximum). The optimal solutions are found for each n, by selecting the row minimum.

Calculated values are presented illustrating this dynamic programming approach of sorting and selecting the minimum values.

Table VI: Optimization Matrix

| n | Site Safety Total of 1 New Unit per Site $TS_1(k)$ | Site Safety Total of 2 New per Site $TS_2(k)$ | Site Safety Total of 3 New per Site $TS_3(k)$ | Site Safety Total of 4 New per Site $TS_4(k)$ |
|---|---|---|---|---|
| 1 | $TS_1(1)$ | | | |
| 2 | $TS_1(1) + TS_1(2)$ | $TS_2(1)$ | | |
| 3 | $TS_1(1) + TS_1(2) + TS_1(3)$ | $TS_2(1) + TS_1(1)$ | $TS_3(1)$ | |
| 4 | $TS_1(1) + TS_1(2) + TS_1(3) + TS_1(4)$ | $TS_2(1) + TS_2(2)$ | $TS_3(1) + TS_1(1)$ | $TS_4(1)$ |

Table VII: Calculations For 150 mi EI *Net Gen Index for Additional n units (500 MWh)

| Variable | Site Index | 150mi EI* 10^3 KWh | Rank | Percent | Variable | Site Index | 150mi EI* 10^3 KWh | Rank | Percent |
|---|---|---|---|---|---|---|---|---|---|
| TS₁ (12) | 7 | 6.40E+10 | 1 | 100.00% | TS₂ (12) | 7 | 7.83E+10 | 1 | 100.00% |
| TS₁ (11) | 3 | 3.09E+10 | 2 | 90.90% | TS₂ (11) | 3 | 3.84E+10 | 2 | 90.90% |
| TS₁ (10) | 2 | 1.72E+10 | 3 | 81.80% | TS₂ (10) | 2 | 2.59E+10 | 3 | 81.80% |
| TS₁ (9) | 11 | 1.60E+10 | 4 | 72.70% | TS₂ (9) | 11 | 2.11E+10 | 4 | 72.70% |
| TS₁ (8) | 10 | 1.31E+10 | 5 | 63.60% | TS₂ (8) | 9 | 2.01E+10 | 5 | 63.60% |
| TS₁ (7) | 9 | 1.01E+10 | 6 | 54.50% | TS₂ (7) | 10 | 1.86E+10 | 6 | 54.50% |
| TS₁ (6) | 6 | 8.01E+09 | 7 | 45.40% | TS₂ (6) | 4 | 1.01E+10 | 7 | 45.40% |

| Variable | Site Index | 10^3 KWh | Rank | Percent | Variable | Site Index | 10^3 KWh | Rank | Percent |
|---|---|---|---|---|---|---|---|---|---|
| $TS_1(5)$ | 4 | 7.52E+09 | 8 | 36.30% | $TS_2(5)$ | 6 | 9.80E+09 | 8 | 36.30% |
| $TS_1(4)$ | 8 | 5.29E+09 | 9 | 27.20% | $TS_2(4)$ | 1 | 7.71E+09 | 9 | 27.20% |
| $TS_1(3)$ | 1 | 5.28E+09 | 10 | 18.10% | $TS_2(3)$ | 8 | 6.65E+09 | 10 | 18.10% |
| $TS_1(2)$ | 5 | 2.64E+09 | 11 | 9.00% | $TS_2(2)$ | 5 | 3.99E+09 | 11 | 9.00% |
| $TS_1(1)$ | 12 | 1.96E+09 | 12 | 0.00% | $TS_2(1)$ | 12 | 3.92E+09 | 12 | 0.00% |

| | $TS_3(k)$ n = 3 | | | | | $TS_4(k)$ n = 4 | | | |
|---|---|---|---|---|---|---|---|---|---|
| | | 150mi EI* | | | | | 150mi EI* | | |
| Variable | Site Index | 10^3 KWh | Rank | Percent | Variable | Site Index | 10^3 KWh | Rank | Percent |
| $TS_3(12)$ | 7 | 9.26E+10 | 1 | 100.00% | $TS_4(12)$ | 7 | 1.07E+11 | 1 | 100.00% |
| $TS_3(11)$ | 3 | 4.60E+10 | 2 | 90.90% | $TS_4(11)$ | 3 | 5.35E+10 | 2 | 90.90% |
| $TS_3(10)$ | 2 | 3.46E+10 | 3 | 81.80% | $TS_4(10)$ | 2 | 4.33E+10 | 3 | 81.80% |
| $TS_3(9)$ | 9 | 3.02E+10 | 4 | 72.70% | $TS_4(9)$ | 9 | 4.03E+10 | 4 | 72.70% |
| $TS_3(8)$ | 11 | 2.62E+10 | 5 | 63.60% | $TS_4(8)$ | 11 | 3.13E+10 | 5 | 63.60% |
| $TS_3(7)$ | 10 | 2.41E+10 | 6 | 54.50% | $TS_4(7)$ | 10 | 2.96E+10 | 6 | 54.50% |
| $TS_3(6)$ | 4 | 1.28E+10 | 7 | 45.40% | $TS_4(6)$ | 4 | 1.54E+10 | 7 | 45.40% |
| $TS_3(5)$ | 6 | 1.16E+10 | 8 | 36.30% | $TS_4(5)$ | 6 | 1.34E+10 | 8 | 36.30% |
| $TS_3(4)$ | 1 | 1.01E+10 | 9 | 27.20% | $TS_4(4)$ | 1 | 1.26E+10 | 9 | 27.20% |
| $TS_3(3)$ | 8 | 8.02E+09 | 10 | 18.10% | $TS_4(3)$ | 8 | 9.39E+09 | 10 | 18.10% |
| $TS_3(2)$ | 12 | 5.88E+09 | 11 | 9.00% | $TS_4(2)$ | 12 | 7.84E+09 | 11 | 9.00% |
| $TS_3(1)$ | 5 | 5.34E+09 | 12 | 0.00% | $TS_4(1)$ | 5 | 6.69E+09 | 12 | 0.00% |

Table VIII: Tabulated Results For Optimums

| n | Site Safety Total of 1 New Unit per Site $TS_1(k)$ | Site Safety Total of 2 New per Site $TS_2(k)$ | Site Safety Total of 3 New per Site $TS_3(k)$ | Site Safety Total of 4 New per Site $TS_4(k)$ |
|---|---|---|---|---|
| 1 | $TS_1(1)$ = Site 12 = 1.96E+09 | | | |
| 2 | $TS_1(1) + TS_1(2)$ = Sites 12 & 5 1.96E+09 + 2.64E+09 = 4.60E+09 | $TS_2(1)$ = Site 12 3.92E+09 | | |
| 3 | $TS_1(1) + TS_1(2) + TS_1(3)$ = Sites 12 & 5 & 1 1.96E+09 + 2.64E+09 + 5.28E+09 = 9.88E+09 | $TS_2(1) + TS_1(1)$ = Sites 12 & 12 = 1.96E+09 3.92E+09 = 5.88E+09 | $TS_3(1)$ = Site 5 5.34E+09 | |

Table IX: Illustrative Minimums Metric Used At the Chosen Sites

| Additional Generation | Site |
|---|---|
| 500 MW | Maine Yankee |
| 1000MW | Maine Yankee |
| 1500MW | Fitzpatrick |
| 2000 MW | Fitzpatrick |

Another optimization was performed to explore the robustness and address two objectives by a weighting. In particular, the objective function representing the site that both maximizes meeting the demand (50 mile population) and minimizes the risk (150 mile Exposure Index) is given by:

Minimize

$$\{TS_n(j)\} = ES(j) + n \cdot \left[ w \cdot POP(j) + (1-w) \cdot NES(j) \right], \quad n = 1,..k$$

Solutions for n = 1,2, 3, 4 new reactors generating 250, 500, 1000, and 2000 MWh and weights w = 0.25, 0.5, and 0.75 were calculated first, then sorted (or ranked), and selections made for the minimums as previously shown. The results are generally the same with Maine Yankee and Fitzpatrick; however, for a 250 MWh size, Shoreham appeared in the optimal set for two sites (Maine Yankee and Shoreham) or one at Fitzpatrick. Maximizing a single objective of meeting demand would select the Indian Point site.

## 5 CONCLUSION:

The menu of tabulated results, calculations, and displays help explore siting alternatives more clearly for decision-making on site population and risk considerations. The array of simple techniques including statistical, optimization, and financial analysis incorporates extensions and new ideas to previous assessments. In particular, statistical z-scores, clustering, and optimization calculations provide useful insights. Additionally, new simple financial derivations have been made using explicit probability models rather than extreme value approaches which assume no particular distributional information. Connections to other complex financial math including diffusional fractional calculus option pricing has been presented

There certainly exists a range of choice for the decision on where to locate new energy facilities. Several sites offer inherent advantages by their location because of population and environmental transport exposure, or other surrogate measures of risk. Further formulation of siting ranking indices by weighted objectives such as of minimizing both close-in and longer range population exposures , or calculating by measures of a statistical nature would reveal their potentially useful application in decision-making . The final decision, of course, would be subjective. However, further research results would help illuminate tradeoffs of significance for an informed decision. A better understanding on the meaning of the objectives of meeting high population density site electricity needs while at the same time keeping accident prevention paramount for reliable operation will help achieve goals of energy independence.

## 6. REFERENCES:

1. Aldrich, D., et. al., Sandia Siting Study, US NUREG/CR-2239, 1982.

2. Aldrich, D.C., *Generic Environmental Impact Statement for License Renewal of Nuclear Plants*, NUREG-1437 Vol. 1, US Nuclear Regulatory Commission, 1996.

3. Beckjord, E., M. Cunningham, and J. Murphy, Probabilistic safety assessment development in the United States 1972–1990, Reliability Engr & System Safety, 39, Issue 2, 1993, pp. 159-170.

4. Bellman, R., *Dynamic Programming*, Princeton University Press, 2010.

5. Bixler, N.E., S.A. Shannon, C.W. Morrow, B.E. Meloche, SECPOP2000 Sector Population, Land Fraction, and Economic Estimation Program, USNRC NUREG/CR-6525, 2000.

6. Blond, R.M.; Burke, R.P.; Margulies, T.S., "Safety Goal Evaluation: Sensitivity Studies," International Meeting on Probabilistic Safety Assessment, SAND-84-2510C; CONF-850206-15, 1985.

7. Campolongo, F., J. Cariboni, and W. Schoutens, The importance of jumps in pricing European options, Reliability Engrg & System Safety, 91, Issues 10-11, October-November 2006, pp. 1148-1154.

8. Cartea, A., and D. del Castillo Negrete, "Fractional Diffusion Models of Option Prices in Markets with Jumps," Birkbeck University of London, Working Papers in Economics & Finance BWPEF 0604, 2006.

9. DuBord, R.M., M.W. Golay and N.C. Rasmussen, "A Probabilistic Risk Assessment-Related Methodology to Support Performance-Based Regulation Within the Nuclear Power Industry," *Nuclear Technology*, 114 169-178, 1996.

10. Dulik, J.D., S.M. Utton and M.W. Golay, "Improved Nuclear Power Plant Operations Through Performance-Based Safety Regulation," in Probabilistic Safety Assessment and Management (PSAM-4), A. Mosleh and R.A. Bari, eds., 385-390, 1998.

11. Everitt B.S., Landau S. and Leese M., Cluster analysis (4th edition). Arnold, London, 2001.

12. Feller, W., *Introduction to Probability Theory and Its Applications* Vol. I, 3 rd Edition, Wiley, 1968.

13. Golay, Michael W., Isi Saragossi, and Jean-Marc Willefert, Comparative Analysis Of United States and French Nuclear Power Plant Siting and Construction Regulatory Policies and their Economic Consequences," Massachusetts Institute of Technology, Energy Laboratory Report No. MIT-EL 77-044-WP, 1977.

14. Greenberg, M. and D. Krueckeberg, "Demographic analysis for nuclear siting: A set of computerized models and a suggestion for improving siting practices," Computers & Operations research, Vol.1, Issues 3-4, 1973, 497-506.

15. Greenberg, M. and D. Krueckeberg, M. Kaltman et al., "Local planning v. national policy, The Town Planning Review, Vol. 57 (3), 1986, 225-237.

16. Hansen, K.F. and M.W. Golay, "Systems Dynamics: An Introduction and Applications to the Nuclear Industry," *Advances in Nuclear Science and Technology*, ed. J. Lewins, M. Becker, 24, 191-221, Plenum Press, NY, 1997.

17. Harbison, S., Safety objectives in nuclear power technology, *Reliability Engineering & System Safety, 31,* Issue 3, 1991, pp. 297-307.

18. Helton, J., J. A. Rollstin, J. L. Sprung, J. D. Johnson, " An exploratory sensitivity study with the MAACS reactor consequence model, *Reliability Engineering & System Safety, 36*, Issue 2, 1992, pp. 137-164.

19. Helton, J., Roger J. Breeding, Calculation of reactor accident safety goals, *Reliability Engrg & System Safety, 39,* Issue 2, 1993, pp. 129-158.

20. Helton, J., J. D. Johnson, M. D. McKay, A. W. Shiver and J. Sprung, "Robustness of an uncertainty and sensitivity analysis of early exposure results with the MACCS reactor consequence model, *Reliability Engrg & System Safety,* 48, Issue 2, 1995, pp. 129-148.

21. Hoel, Paul G., *Introduction to Mathematical Statistics*, 5-th Edition, Wiley-Interscience, 1984.

22. Ito, K., On stochastic differential equations. *Memoirs, American Mathematical Society,* 4, 1951, pp. 1–51.

23. Li, H., Apostolakis, G.E., Gifun, J., VanSchalkwyk, W., Leite, S., and Barber, D., "Ranking the Risks from Multiple Hazards in a Small Community," *Risk Analysis,* 29: 2009, 438-456.

24. Mainardi, F., R. Gorenflo, and A. Vivoli, "Renewal Processes of Mittag-Leffler and Wright type, "Fractional Calculus and Applied Analysis, 8, 1, pp. 7-38, 2005.

25. Margulies, T., "Evaluation and Comparison of High Population Density Sites, Johns Hopkins University Applied Physics Laboratory, Report PPSE-T-12, NTIS PB81-196719, 1979.

26. Margulies, T., T. Eagles, J. Cohon, "Multi-Objective Regional Energy Location Cost Versus People Proximity Trade-Offs", *Trans. Am. Nuc. Soc.*, Vol. 38, 1981, 117.

27. Margulies, T., et. al., *The Development of Severe Reactor Accident Source Terms: 1957-1981*, U.S.

Nuclear Regulatory Commission, NUREG-0773, 1981.

28. Margulies, T., T. Eagles, J. Cohon, "Analysis of Reactor Siting Policy Using Multi-Objective Programming," Proc. Fifth International Meeting on Multiple Criteria Decision-making, Edited by Pierre Hanson, Belgium, 1982.

29. Margulies, T. and R. Blond, "Variability of Site Reactor Risk," *Journal of Risk Analysis*, Vol. 4, No. 2, 1984, 89.

30. Margulies, T., "Risk Optimization: Siting of Nuclear Power Electricity Generating Units," *Reliability Engineering and System Safety*, vol. 86/13, 2004.

31. Margulies, T., "Simple Cost Risk-Benefit Calculation: Nuclear Plant Back-fit Analysis, "Reliability Engineering and System Safety, vol. 86/13, 2004.

32. Merton, R., *Continuous Time Finance*, Blackwell, 1st Edition, 1990.

33. Munson, C. G., A. J. Kigler, 2010, "Insights from Siting New Nuclear Plants in the Central And Eastern United States, Presentation in Prague, Czech.

34. Otway, H. and R. Erdmann, "Reactor siting and design from a risk viewpoint,"Nucl. Engr. and Design, 13, Issue 2, August 1970, Pages 365-376.

35. Ramsay, W., 1977, *Environ. Sci. Technol.*, 11 (3), pp 238–243.

36. Rasmussen, Professor Norman C.; et al., Reactor Safety Study: An Assessment of Accident Risks in U.S. Commercial Nuclear Power Plants, Executive Summary, WASH-1400 (NUREG- 75/014); Reactor Safety Study: An Assessment of Accident Risks in U.S. Commercial Nuclear Power Plants (NUREG-75/014), Appendix VI, 1975.

37. Rodwell (Project Manager), E., *Siting Guide: Site Selection and Evaluation Criteria for an Early Site Permit Application*, 1006878, Final Report, Electric Power Research Institute, 2002.

38. Tait, N., The use of probability in engineering design—an historical survey, *Reliability Engineering & System Safety, 40, Issue 2, 1993, pp. 119-132.*

39. Sorensen, J., "Safety culture: a survey of the state-of-the-art," Reliability Engnrg & System Safety," 76, Issue 2, May 2002, pp. 189-204.

40. United Kingdom Health and Safety Directorate, Land Use Planning and Siting of Nuclear Installations in the United Kingdom.

41. U.S. Nuclear Regulatory Commission, 'General Site Suitability Criteria for Nuclear Power

Stations', Regulatory Guide 4.7, Revision 2, 1998.

42. Ward J.H., "Hierarchical grouping to optimize an objective function," Journal of the American Statistical Association, 58,1963, 238-244.

43. Weil, R. and G. E. Apostolakis, A methodology for the prioritization of operating Experience in nuclear power plants, *Reliability Engrng & System Safety*,74 Issue 1, October 2001, pp. 23-42 .

44. Yambert, M., and M. Linn, *Assessment of the Exposure Index as a Means of Predicting Potential Consequences Associated with Nuclear Power Plant Accidents*, ORNL/NRC/LTR-92/10/RI, Oak Ridge National Laboratory, Oak Ridge, Tennessee, 1992.

45. Zio, E., Reliability engineering: Old problems and new challenges, Reliability Engineering and System Safety, 94, Issue 2, February 2009, pp. 125-141

Appendix A1. Notes on Fractional Calculus

These notes introduce some definitions or algorithms for generalizing integer calculus to fractional order [Hosking 1981; Oldham & Spanier 2006 ;Polubny (1999, 2000, 2002); Diethelm, et al. 2005].

The calculus of functions for a fractional order of differentiation and integration has been traditionally derived from several lines of investigation, including for example: the Grunwald-Letnikov definition which reduces to the Riemann sum or backward difference limit; the Riemann-Louiville method of iterated integrals also derives a differ-integral by generalizing from the first integral of the function of f to multiple integrations; as well as, a Cauchy integral formulation for a closed contour surrounding the point enclosing an analytic region. The literature cited contains excellent presentations of these and several transform modifications and diffusion equation solutions [Negrete 2006 ; Yuste 2006 ].

Using a differencing approach [Hosking 1981] applied in statistical time series analysis for autoregressive integrated moving averages, where the B operator implements a translational or shifting, where the B operator produces a translation or shifting,

$$Bx_t = x_{t-1}$$

backwards say in time t,

$$\nabla^d = (1-B)^d = \sum_{i=0}^{n \to \infty} \binom{d}{i}(-B)^{n-i} = \sum_{i=0}^{n \to \infty} (-1)^{n-i}\binom{d}{i}(B)^{n-i},$$

$$= 1 - dB - 0.5d(1-d)B^2 - \frac{1}{6}d(1-d)(2-d)B^3 - \ldots$$

If the independent variable or input $x(\tau)$ is viewed as an infinite sum of weighted and shifted impulses, a dependent variable or output $y(t)$ with these inputs is a superposition of their responses as in a filtering of data or convolution given by.

$$y(t) = \int x(\tau)h(t-\tau)d\tau = x*h(t) = h*x(t) =$$

$$\int x(t-\tau)h(t)d\tau = x*h(t)$$

The convolution integral is defined as $y(t) = \int_{-\tau} x(\tau)d\tau$ or in discrete form,

$$x[n] = \sum_{j=0}^{\infty} x[j]\delta[n-j] \text{ and} \qquad y[n] = \sum_{j=0}^{\infty} x[j]h[n-j]$$

Other similarities appear to the autoregressive integrative moving average calculations of time series analysis for data exhibiting memory.

References

A1) Aoun, M., R. Malti, F. Levron, and A. Oustaloup, "Synthesis of fractional Laguerre basis for system approximation," Automatica Preprint.

A2) Diethelm, K., et al., 2005, "Algorithms for the fractional calculus: A selection of numerical methods, Comput. Methods Applied Mech Engrg, 264-274.

A3) Hosking, J., Fractional Differencing, Biometrika, 68, No. 1, 1981, pp. 165-176.

A4) Negrete, D., Fractional diffusion models of non-local transport, Phys. Of Plasmas, 2006.

A5) Oldham, K.B. and Jerome Spanier, 2006, *The Fractional Calculus*, Dover.

A6) Podlubny, I.,1999, *Fractional Differential equations*, Mathematics in Science and Engineering (Volume 198), Academic Press.

A7) Polubny, I., 2000, "Matrix Approach to Discrete Fractional calculus," Fractional Calculus and Applied Analysis," Vol. 3, 4, 359-386.

A8) Podlubny, I., 2002, "Geometric and Physical Interpretation of Fractional Integration and Fractional Differentiation," Fractional Calculus and Applied Analysis, 5, 367-386.

A9) Yuste, S.B., 2006, "Weighted average finite difference methods for fractional diffusion equations," J. Comput Phys. 216, , 264-274.

## Multiple Purpose Planning and Analysis with Population

A free market system may for the best goods and services to prosper is viewed as being directed by "an invisible hand" toward the public good as espoused by Adam Smith (1937). Moral development has been described as non-linear and multi-dimensional compared to a uni-directional progression. An amoral organization pursues profit at any cost. Following the letter and spirit of the law has always been the challenge. Ethical decisions can be made on long-term economic investments merging social and responsible investments with ethics as a goal. Decisions to help the welfare of society as a whole are best when made from examining all the alternatives and what each alternative action would mean taking into account multiple conflicting aims or goals.

Multiple objectives or purposes for planning and policy analysis has been an agglomeration of achievements in psychology, economics, and operations research. Modern approaches stem from the words and ideas expressed in the Flood Control Act of 1938. That is, costs [in the Act they were related to water resource projects such as dams] must be taken into account to whomsoever they may accrue. This kicked-off a cost-benefit analysis thinking in all areas of planning both domestic and military. The costs to whom thinking that accompanies multiple goals or aims are explicitly accounted for in a multiple purpose planning framework. Decisions ideally would be based on the best available information and analysis to make informed decisions among a full range of choice of alternatives. In general, information can be categorized as nominal, ordinal, or cardinal. That is, nominal information represents a descriptive approach, while ordinal information represents ranks, and cardinal data are numerical, usually with real numbers. A mix of all three kinds of information comprises documentation accompanying safety and environmental impact reports.

An initial allocation of goods such that a different allocation that makes at least one individual better off without making any other individual worse off is called a Pareto improvement to credit its Italian economist inventor. An allocation is defined as "Pareto efficient" or "Pareto optimal" when no further Pareto improvements can be made.

Notions of pareto optimality for offering the best in a single objective optimization of a welfare state are generalized. The original idea of optimality was that an individual's worth or value is maximized (or minimized in the model formulation) without taking away from another individual in society. Surrogate worths by multiple attributes or mathematical programming techniques are displayed as trade-off s in meaningful ways reagrding the decisonmakers' aims. The intent is to provide the best understanding of what each choice means among the objectives having meaning for a public expendature.

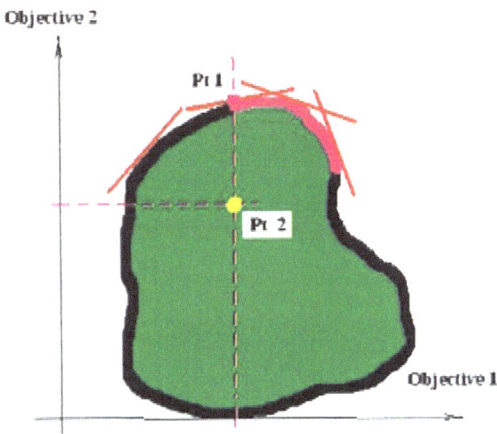

Macroscopic goals for society include economic stabilization and structuring with institutional policies emphasizing balanced perspectives on energy generation, environmental protection, as well as optimizing alternative supplies. Goals include minimizing population risks from hazardous facilities and from transportation of hazardous materials to sustains a secure homeland and future. Efforts are continuously needed to monitor pipelines especially near populated or sensitive environmental areas, evaluate nuclear plant performance, and implement procedures and design changes to minimize chemical industry hazards. Slowing shipping of tankers of oil from overseas travel across oceans to save costs, as well as quieting the sea environment from engine and fluid-structure noises has benefits for sea mammal populations and their communications has been proposed.

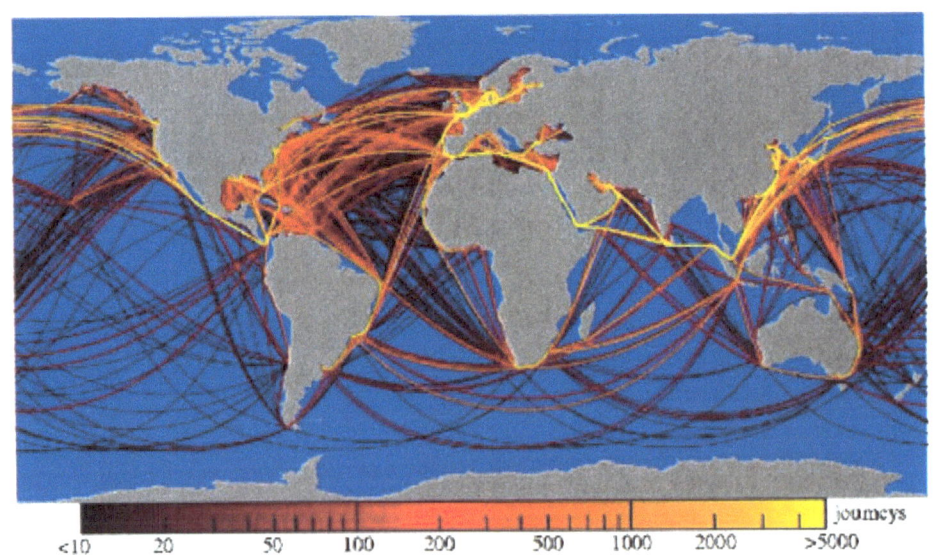

Major Marine Shipping Routes: [Refs: Image: Bernd Blasius/J. Roy. Soc Interface 19 Jan 2010 pp 1-11 ; Kaluza, Andrea Kolzsch, Michael T. Gastner, and Bernd Blasius, **The complex network of global cargo ship movements**]

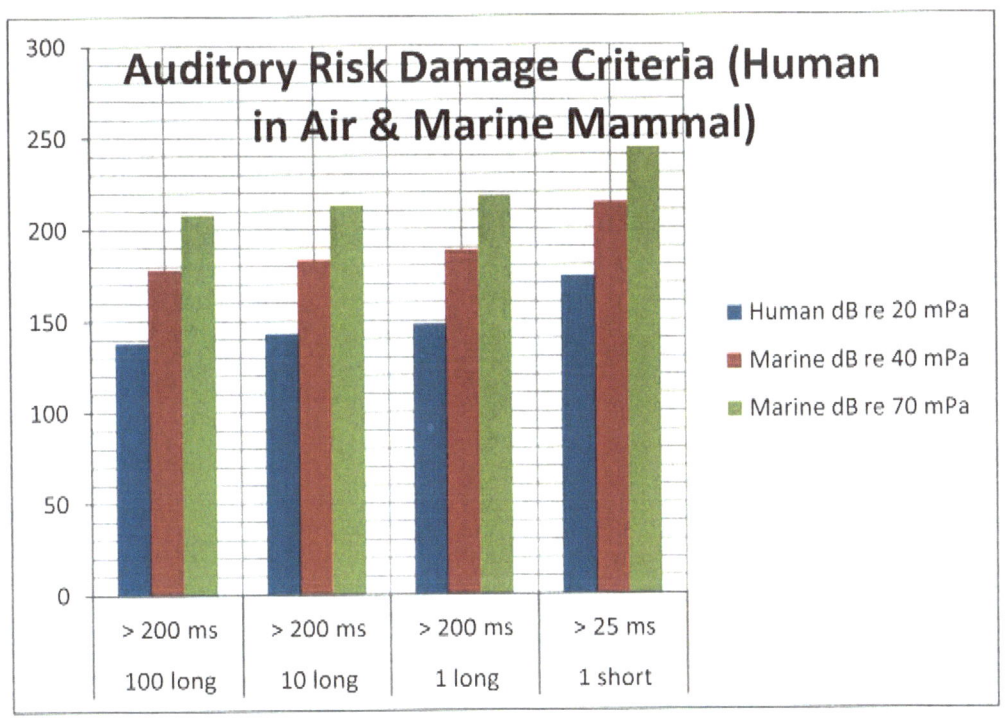

The reference intensities used to compute decibel (dB) units are different in water and air. A sound wave with a pressure of 1 microPascal (µPa) is a reference used for underwater sound. In air, use of a sound wave with a pressure of 20 µPa as the reference is made. This value for sound pressure in air of 20 µPa corresponds to sounds at a frequency of 1000 Hz that can just be heard by humans. Intensity levels are other measures of sound (the pressure squared for a plane small amplitude wave) with generally vector or directional information. The dB difference between air and underwater sound pressures is elaborated further.

$dB_w = 10 \cdot log\left(\frac{p^2}{p_w^2}\right)$ or equivalently $p = p_w \cdot 10^{\frac{dB_w}{20}}$; $dB_w = 20 \cdot log\left(\frac{p}{p_w}\right) = 20 \cdot log(p) - 20 \cdot log(p_w)$ and $dB_A = 10 \cdot log\left(\frac{p^2}{p_A^2}\right) = 20 \cdot log(p) - 20 \cdot log(p_A)$. Therefore

$dB_w - dB_A = -20 \cdot log(p_w) + 20 \cdot log(p_A)$

Computing the difference    -20*Log10(1*10^-6)-(-20*Log10(20*10^-6)) =

26.02059991

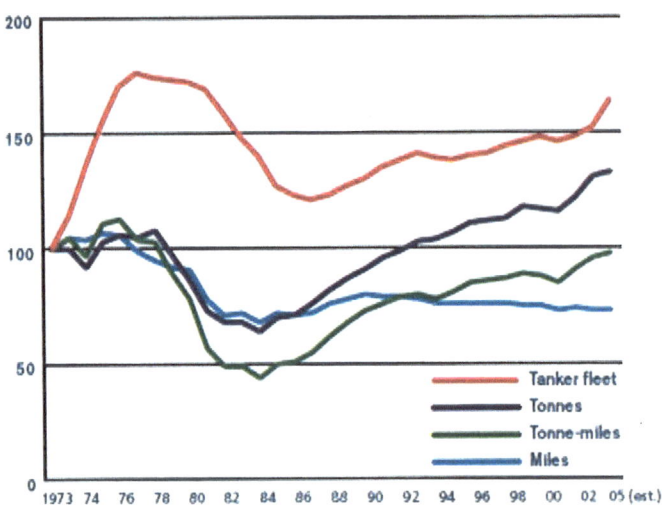

Source: Intertanko Tanker Facts 2006,
http://www.intertanko.com/about/annualreports/2005/5.html

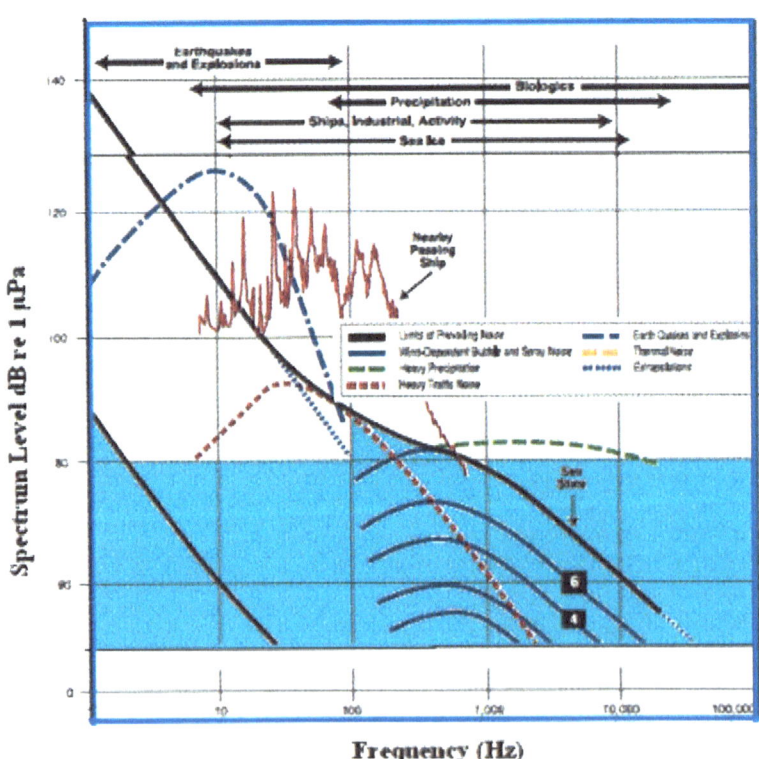

Marine Container Ship Transportation Cost Algebra

In response to fuel costs and uncertainties, analysts have been striving for a better understanding of the number of ships, their speed, in addition to frequency of port visits has been attempted with algebra.

$$T_r[days] = \sum_{i=1}^{n} T_{p,i}[days] + \frac{D[nm]}{V[knots]\cdot\frac{24hrs}{day}}, \quad knots \equiv \frac{nm}{hr} \tag{1}$$

The letters denote variable numbers as defined by,

$T_r$ Total round trip voyage time; $T_{p,i}$ Time in-port I; $D$ Total distance traveled; $V$ Vessel speed

$$T_r \leq \frac{7 \cdot S}{F} \tag{2}$$

$F$ frequency of service per week in each port call; $S$ Number of ships in liner service

Solving the above equations as a system of algebraic equations obtains an inequality,

$$V \geq \frac{D}{\left(168\frac{S}{F} - 24(\sum_{i=1}^{n} T_{p,i})\right)} \tag{3}$$

The marine speed unit of knot is equal to one nautical mile (1.852 km) per hour which is approximately 1.151 mph. Historically, vessel travelling at 1 knot along a meridian travels one minute of geographic latitude in one hour and a chip log was used for measurement.

The Gray and Greeley model for predicting blade rate noise for a merchant fleet typically uses the following equation:

$$P_{rms,d} = \frac{\sqrt{2}\pi^2 \rho \cdot d \cdot V_{max} \cdot f^3}{r \cdot c} sin\theta$$

where $P_{rms,d}$ is the rms dipole pressure; $\rho$ is the fluid density, $f$ is frequency, $d$ is the source depth, $V_{max}$ is the maximum size of the cavity volume, u is the radiation angle (0=horizontal), $r$ is distance to a field point, and $c$ is the sound speed. Using empirical cavity size data, the dipole source pressure level of blade rate is given by

$$P = 146 + 10 \cdot log_{10}(f^6 d^2 V_{max}^2)$$

where the level is measured in dB re 1 μPa at 1 m.

The predicted dipole source level is 171 dB re 1 μPa at 1 m at 140 rpm. A monopole level can be computed by summing -10 log(2kd)², or 17 dB in this example, obtaining 188 dB re 1μPa at 1 m.

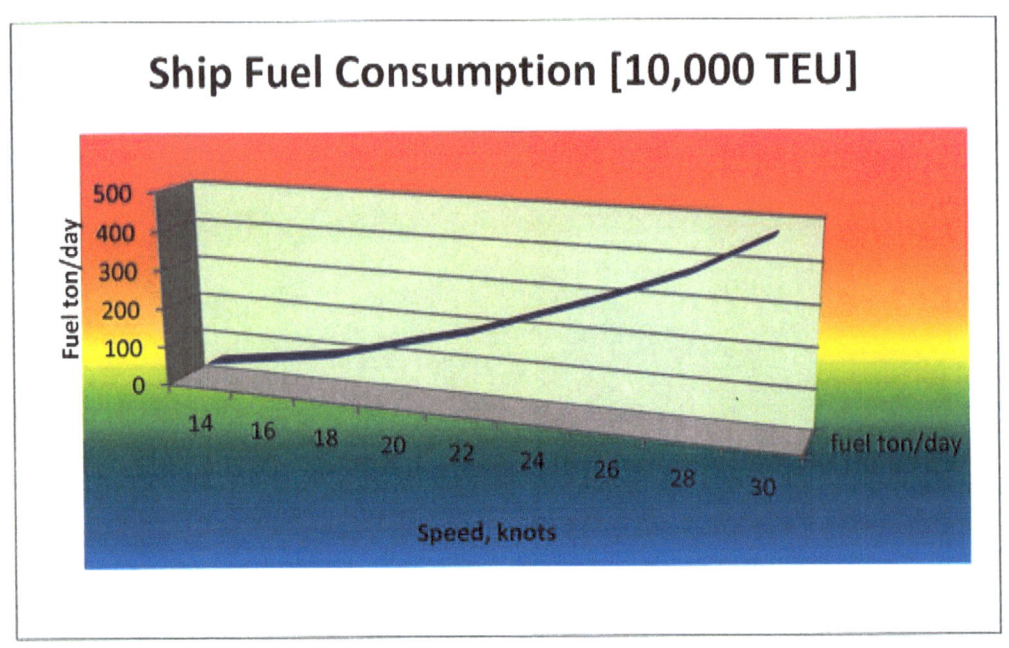

For illustration of the formula in equation (3) and using the following calculations were created with Excel. twenty equivalent feet unit.

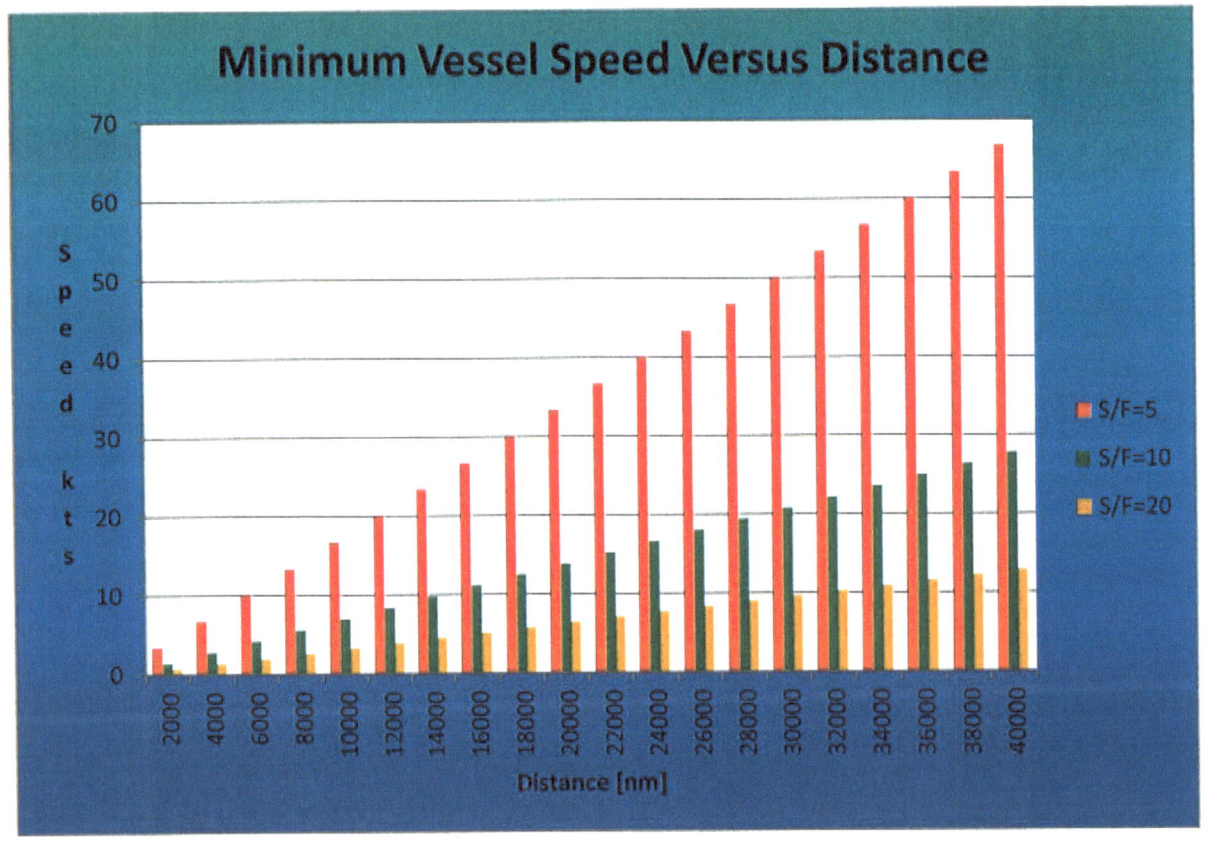

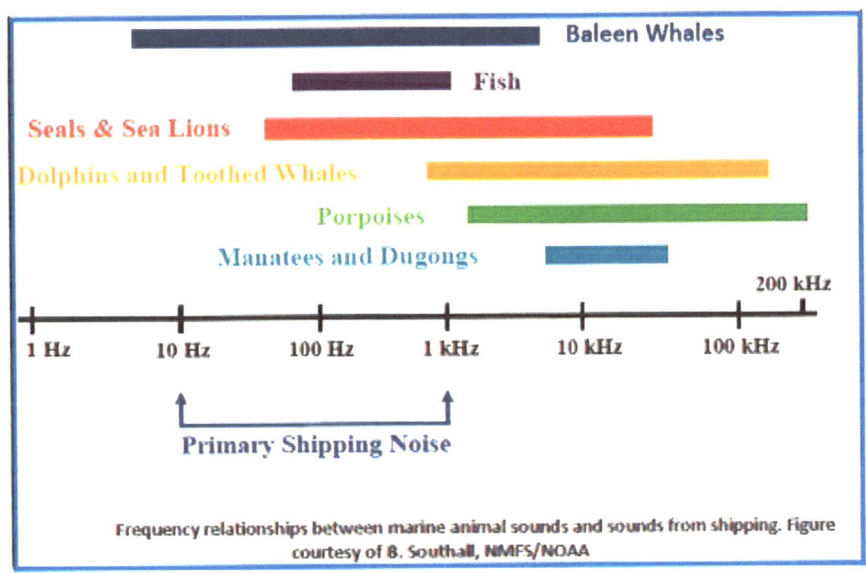

Frequency relationships between marine animal sounds and sounds from shipping. Figure courtesy of B. Southall, NMFS/NOAA

The achievement of low speed shipping appears more favorable for long transits when the ratio of number of ships to frequency is increased. The reliability of the shipping service may override response time factors for some commodities as well as achieve lower ocean noise levels.

The tools of multiple objective decision-making have been successfully applied to energy systems such as commercial nuclear power. These policy and planning applications in both transportation of spent fuel to an interim facility and the selection of nuclear generating sites.

Several Nuclear Issues Regarding Back-fits to Power Reactors and Their Spent Fuel

| BWR | PWR | New |
|---|---|---|
| Automatic Depressurization | Event-V Check Valve | Advanced Reactors |
| Supplemental Containment/ Iodine Removal/ | | Pebble Bed Design |
| Mark I, II Modification (Filtering, Hardening) | | Partitioning/Transmutation of Spent fuel; Vitrification/Storage |

The power system would benefit from, for example, ex-vessel *core catchers* in the event of a meltdown to prevent excessive loading of containment volumes and cool with water and to confine the molten material as a new fourth barrier of defense-in-depth especially for plants with uncertain risks. The menu of additional improvements could include an automatic leak-monitoring system of containment for unintended openings, as well as implementation of reliable lubricating systems and fan coolers that support critical plant systems during an accident. Further design alternatives were assessed for the Westinghouse Advanced design.

Transportation policy and planning analysis have provided insights on answering the following questions for waste management of the spent fuel from reactor sites: Which Away-From-Reactor facility should be opened? Which AFR should each reactor ship its spent fuel? Along which routes should the spent fuel be shipped? Mathematical models with computer implementation answered all three questions simultaneously with multiple objectives. The two objectives modeled were measures of transportation risk, that is, ton-miles and people-ton-miles along a widened highway corridor . These objectives were surrogates of the cost of transportation (measured by the amount of fuel shipped and distance traveled) and the conditional risk (or consequences given a major accidental release affecting the population) in transportation environment zones ( within a distance away from the highways). These objectives were technically supportable as well as easily communicated by their units. Because of the many possible solutions to this optimization problem this multi-objective approach appeared attractive to display the range of choice of the non-dominated, or non-inferior set of solutions. The choice of a final processing or repository location could be discussed and other scenarios analyzed with further model development.

There certainly exists a range of choice for decisions. The final decision, of course, would be subjective. However, further research results would help illuminate tradeoffs of significance for an informed decision. A better understanding on the meaning of the objectives, for example of meeting high population density site electricity needs while at the same time keeping accident prevention paramount for reliable operation will help build trust and achieve goals of energy independence.

Homeland safety enhancements based on the saving of lives would build new trust in a technology culture and government. Safe futures and more jobs with the present commercial nuclear plant technology would be seen with a partitioning and transmutation of wastes policy. Smaller, less dangerous amounts would be placed in geologic burial sites for smaller time-frame. The transmutation and partitioning technology has been demonstrated in France with a breeder reactor. The idea of self-realization involving a transmutation of psychological toxics and poisons in the body to purify oneself is a Hawaiian spiritual practice. The alternative of a vitrified or glass-made waste form for a more stable material to bury underground in a deep geologic storage system is achievable as well to minimize risks to surrounding populations.

References

Arveson, Paul T., David J. Vendittis, Radiated Noise Characteristics of a modern cargo ship, J. Acoust. Soc. Am., Vol. 107, No. 1, January 2000, 118-129.

Cohon, Jared L, Multi-Objective Programming and Planning, Dover (2004).

Fischer, Manfred, The Severe Accident Mitigation Concept and the Design Measures for Core Melt Retention of the European Pressurized Reactor (EPR), Nuclear Engineering and Design 230 (2004) 169–180.

Gray, L. M. and D. S. Greeley, "Source level model for propeller blade rate radiation for the world's merchant fleet," J. Acoust. Soc. Am. **67**, 516–522, 1980.
Hummels, David, Transportation Costs and International Trade in the Second Era of Globalization, *Journal of Economic Perspectives—Volume 21, Number 3—Summer 2007—Pages 131–154.*

McKenna, Megan F., Donald Ross, Sean M. Wiggins, and John A. Hildebrand, Underwater radiated noise from modern commercial ships, J. Acoust. Soc. Am. Volume 131, Issue 1, pp. 92-103 (2012).

Mazzuca, Lori Lee, Potential Effects of Low Frequency Sound (LFS) from
Commercial Vessels on Large Whales, Master's Program, University of Washington
2001.

Margulies, Timothy S., *Location Analysis, Movements, and Renewal:
Mathematical Safety-Risk and Dynamics*, ISBN-13: 978-1477415269.

Notteboom, The E. and Bert Vernimmen, The effect of high fuel costs on liner service configuration in container shipping, Journal of Transport Geography (via internet: in press).

Overview of the impacts of anthropogenic underwater sound in the marine environment, OSPAR Commission Conference, 2009.

Revelle, Charles S., Earl Whitlatch, Jeff Wright, Civil and Environmental Systems Engineering, Prentice Hall; 2 edition (August 25, 2003).

Smith, Adam, *The Wealth of Nations* [1776], The Modern Library, Random House 1937.

Weilgart, L.S., The impacts of anthropologic ocean noise on cetaceans and implications for management, NRC Resaech website, cjz.nrc.ca, 2007.
Talley, Wayne K., Ocean Container Shipping: Impacts of a Technological Improvement, Journal of Economic Issues Vol. XXXIV No. 4 December 2000.

Wright, Andrew, International Workshop on Shipping Noise and Marine Mammals, Okeanos Foundation of the Sea, Hamburg, Germany April 21-24, 2008.

## On Nuclear Safety Risk of Hope Creek Site:

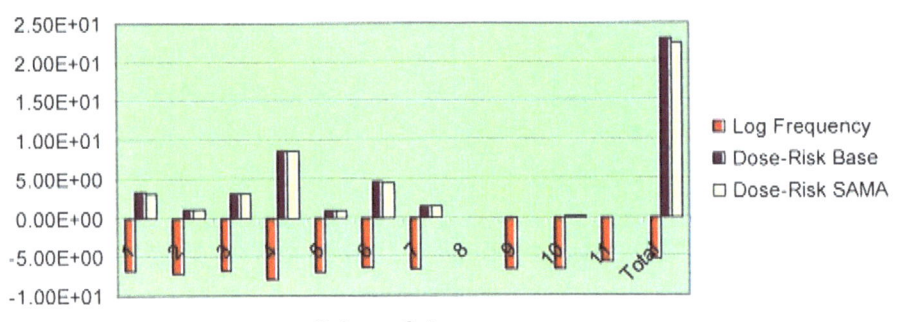

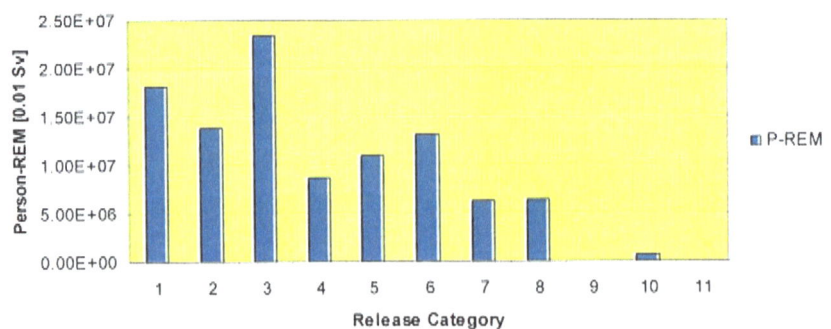

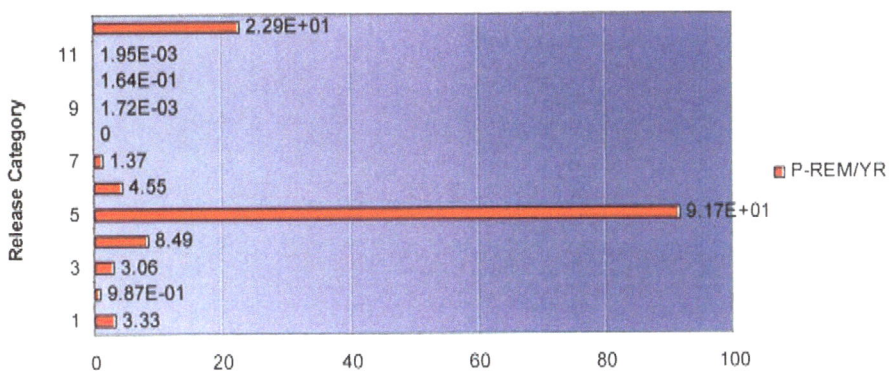

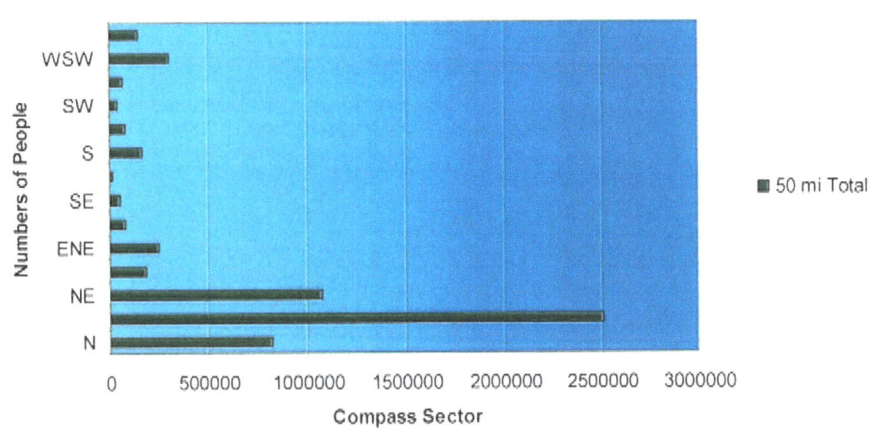

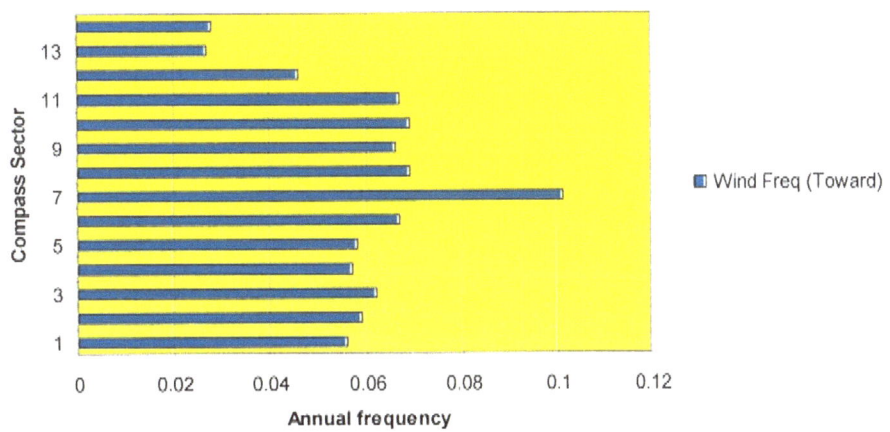

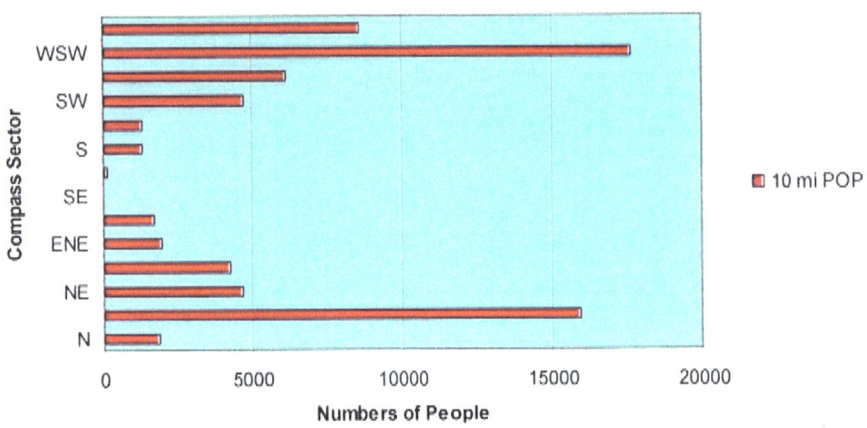

**Hope Creek Ten Mile Population Estimates**

## Chosen Severe Accident Mitigation Alternatives (SAMA's)

## For Proposed Implementation

| SAMA (Severe Accident Mitigation Alternative) |
|---|
| Remove ADS Inhibit from Non-ATWS Emergency Operating Procedures |
| Install Back-Up Air Compressor to supply AOV's |
| Provide Procedural Guidance To Cross-Tie RHR Trains |
| Procedural Guidance to Use Low Pressure Pump for Non-Security Events |
| Replace A Supply fan with Different Design |
| Replace A Supply fan with Different Design |
| Procedural Guidance for Partial transfer of Control Functions to Remote Shutdown Panel |
| Relocate, Minimize, and/or Eliminate Electrical Heaters in Electrical Access Room |
| Procedural Guidance to Bypass RCIC Turbine Trip |

| Category | Expected prem | no wind freq | Unc factor x 25 | $M, 15-yr (4%) | $M,30-yr (4%) |
|---|---|---|---|---|---|
| 1 | 3.3300 | 53.28 | 1332 | 256.67 | 844.506485 |
| 2 | 0.9800 | 15.68 | 392 | 75.536 | 248.533 |
| 3 | 3.0600 | 48.96 | 1224 | 235.86 | 776.0329 |
| 4 | 8.4900 | 135.84 | 3396 | 654.396 | 2153.111 |
| 5 | 91.7000 | 1467.2 | 36680 | 7068.0955 | 23255.629 |
| 6 | 4.5500 | 72.8 | 1820 | 350.707 | 1153.90525 |
| 7 | 1.3700 | 21.92 | 548 | 105.597 | 347.4396 |
| 8 | 0.0000 | 0 | 0 | 0 | 0 |
| 9 | 0.0017 | 0.02752 | 0.688 | 0.13257 | 0.4362 |
| 10 | 0.1640 | 2.624 | 65.6 | 12.64 | 41.5913 |
| 11 | 0.0020 | 0.0312 | 0.78 | 0.1503 | 0.4945 |
| 12 | 22.9000 | 366.4 | 9160 | 1765.096 | 5807.567 |

## Potentially Riskier U.S. Nuclear Plants

There are 23 nuclear reactors in 16 power plants in the United States that have the same reactor and containment design as the Fukushima Daiichi Plant in Japan, where the authorities have been struggling to avert a meltdown. The design, General Electric's Mark 1, is thought to be more susceptible to failure in an emergency than competing designs.

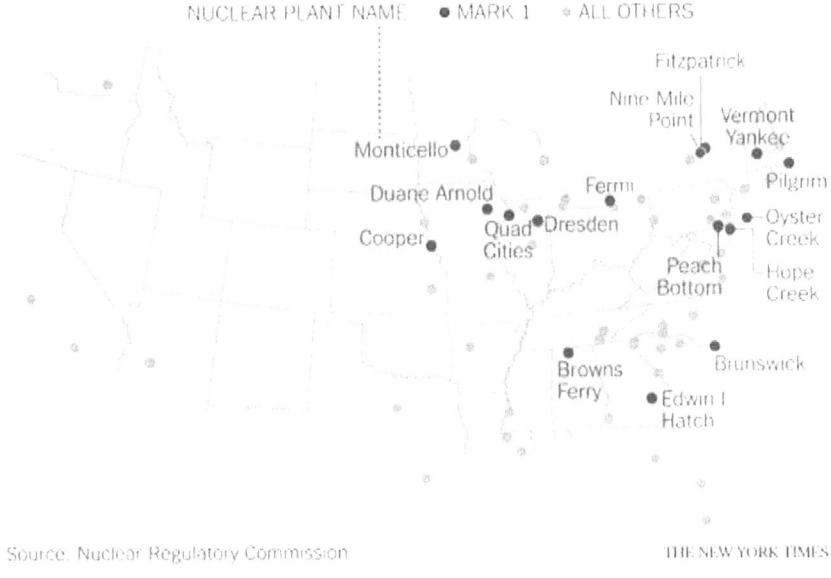

Source: Nuclear Regulatory Commission  THE NEW YORK TIMES

There is new motivation to address unfinished engineering assessment of alternatives to reduce offsite consequences of a large radiological release.

Table: Comparison of Estimated Benefits from Averted Offsite Costs [Ref: Environmental Assessment By the U.S. Regulatory Commission Relating To the Certification of the AP600 Standard Plant Design Docket No. 52-003]

| Design Alternative | Estimated Capital Cost |
|---|---|

| | |
|---|---:|
| Upgrade Chemical & Volume Control System | 1,500,000.00 |
| Filtered Containment Vent | 5,000,000.00 |
| Self-Actuating Containment Isolation Valves | 33,000.00 |
| Passive Safety Grade In-Containment Sprays | 3,900,000.00 |
| Active High-Pressure Safety Injection System | 20,000,000.00 |
| Steam Generator Shell-Side Heat Removal System | 1,300,000.00 |
| Direct Steam Generator Safety & Relief Valve | 620,000.00 |
| Increased Steam Generator Pressure Capability | 8,200,000.00 |
| Secondary containment Filtered Ventilation | 2,200,000.00 |
| Diverse IRWST Injection Valves | 570,000.00 |
| Diverse Containment Recirculation Valves | 150,000.00 |
| Ex-Vessel Core Catcher | 1,660,000,000.00 |
| High Pressure Containment Design | 50,000,000.00 |
| Increased Reliability of Diverse Actuation System | 470,000.00 |

Furthermore, a research update and information on masks and pharmaceutical bio-protectors to individuals offsite could also help the risk decision-making on radiological protections in an accident. Stockpiles could be made for emergency workers and special population groups not likely evacuated until a relocation phase response has begun. These may also serve a dual purpose in being beneficial for civilian defense during military or terrorist acts.

## On Illinois Nuclear Siting Perspectives

Some Illinois nuclear siting perspectives with population measures for near term and long-term risk are given with explorations of alternatives for decision-making. The close-in population is weighted by wind direction frequency as a surrogate risk measure of early fatalities conditional on a severe accident while the fifty-mile population reflects the longer time-frame effects such as latent cancers that could result from a severe accidental release of radioactivity. The displayed graph shows estimated quantified results for these goals. Minimizing both goals requires small populations close-in, as well as far away from the power reactor.

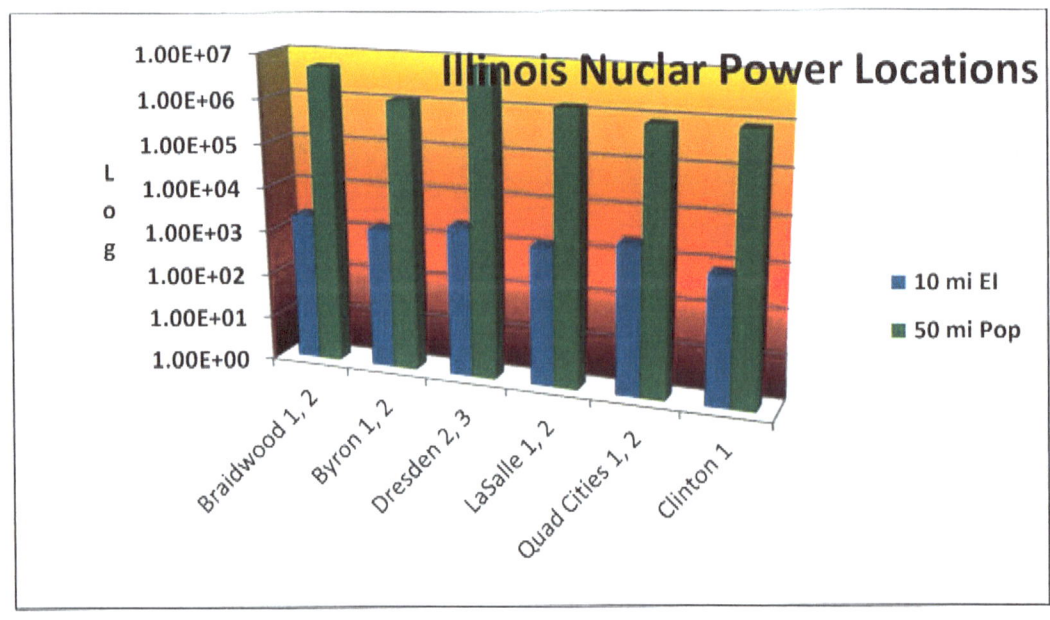

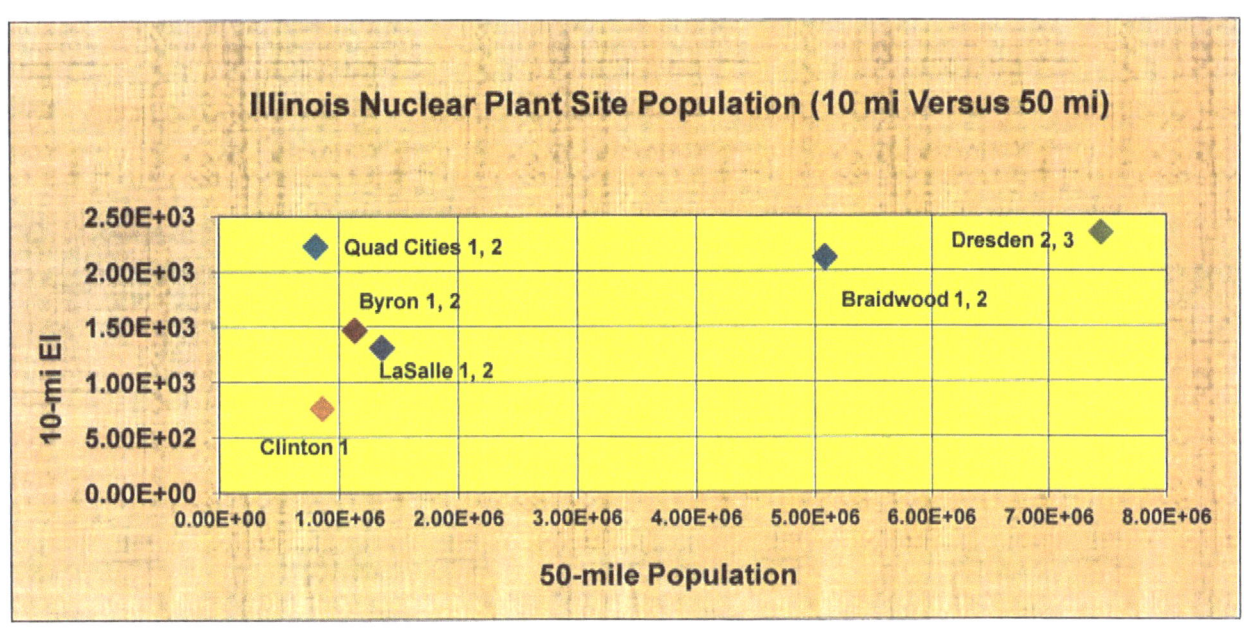

## Pre-Calculus Financial Math With Energy Applications

The simple interest on a principal $P$ ( deposit or investment ) is that amount multiplied by interest rate ( say annual rate, per year) as well as multiplied by the time ( say in years ).

$$I = P \cdot r \cdot t \qquad (1)$$

Of course, the total amount consists of the initial deposit P added to the interest $I$:

$$T = P + I = P + P \cdot r \cdot t = P(1 + r \cdot t)$$

II. Compound Interest Formula

If a principal P has been invested at an interest rate $r$, compounded n times per year, in $t$ years it has grown to a future amount $A$.

$$A = P\left(1 + \frac{r}{n}\right)^{nt}$$

Geometric Series for Periodic Payments (Simple Annuity Calculation)

Consider the summation

$$a + ar + ar^2 + ar^3 + \ldots ar^{n-1}$$

Which consists of n terms. This particular series of terms has its ratios of adjacent terms equal to a constant; that is, consider the k-th and (k-1)st terms in the series, $\dfrac{a \cdot r^k}{a \cdot r^{k-1}} = r$

Denote by $S_n$ the total from summing the n terms,

$$S_n = a + ar + ar^2 + ar^3 + \ldots ar^{n-1}$$

This sum can be found by multiplying each term in the sum by r, and then differencing the latter calculated series with that formerly defined as follows,

$$S_n = a + ar + ar^2 + ar^3 + \ldots ar^{n-1}$$
$$rS_n = ar + ar^2 + ar^3 + \ldots ar^{n-1} + ar^n$$

So that
$$S_n - rS_n = a - ar^n$$

By factoring using the distributive property, $S_n(1-r) = a(1-r^n)$

Dividing by $(1-r)$ yields, $S_n = \dfrac{a - ar^n}{(1-r)} = \dfrac{a(1-r^n)}{(1-r)}$

An Alternative Energy Comparison

$I_1$ Purchase/Installation Solar Electric  $ 29.7 K

$I_2$ Purchase/Installation Electric Service $ 5.0 K

$O_1$ Operating Solar Electric Cost $a_1$ =$ 0.15 K

$O_2$ Operating Electric Service $a_2$ =$ 1.1 K

The problem is formulated for present value calculation; the operating/maintenance time stream of costs is calculated by a present value calculation as well using the first year as the base case year.

$I_1$ and $I_2$ are purchase/installation costs in a present valuation, while maintenance and repair costs occur over time and are calculated on the same present value basis by,

$$O_1 = \sum_{n=0}^{N-1} \frac{O_{1,n}}{(1+i)^n}, \quad O_2 = \sum_{n=0}^{N-1} \frac{O_{2,n}}{(1+i)^n}$$

The interest rate may be made a variable, as well as the time t for calculation of the time (n = N years) when the expenditures would be the same. This "break-even" perspective on the alternative choices may help in the decision-making, for example, of savings over the lifetime of the alternatives.

Equating the total system costs (without decommissioning or waste considerations here) obtains,

$$I_1 + \sum_{n=0}^{N-1} \frac{O_{1,n}}{(1+i)^n} = I_2 + \sum_{n=0}^{N-1} \frac{O_{2,n}}{(1+i)^n}$$

This may be solved for the variable t to find when the costs (installation and operating) are the same. Since

$$I_1 - I_2 = \sum_{n=0}^{N-1} \frac{O_{2,n} - O_{1,n}}{(1+i)^n}$$

where for constant operating/maintenance/repair costs year to year,

define $\quad OC \equiv O_{2,n} - O_{1,n}, \quad$ and $\quad \dfrac{I_1 - I_2}{OC} \equiv \alpha$

Simplifying notation for the equation to be solved,

$$\sum_{n=0}^{N-1} \frac{1}{(1+i)^n} - \alpha = 0$$

One may define a new variable,

$z = \dfrac{1}{(1+i)}, \quad$ then $\quad \sum_{n=0}^{N-1} z^n - \alpha = 0 \quad$ or $\quad \sum_{n=0}^{N-1} z^n = \alpha$

$$S_N \equiv \frac{z - z^N}{(1-z)}, \quad z - z^N = (1-z)\alpha, \quad z - (1-z)\alpha = z^N$$

$$\frac{\log[z - (1-z)\alpha]}{\log(z)} = N$$

Relating nominal and real discounts (without inflation)

$$\hat{z}^{-1} = (z^{-1}(1+e) - 1)$$

with inflation factor e.

## Alternative ENERGY Comparison

| Option | I, [K$] | a, [K$] |
|--------|---------|---------|
| 1 | 2.97E+01 | 1.50E-01 |
| 2 | 5.00E+00 | 1.10E+00 |

I = Installation Cost     α  2.60E+01

| rate % | rate*0.01 | z(i) | Uses: PV(i) |
|--------|-----------|----------|-------------|
| 1 | 0.01 | 0.990099 | 3.13E+01 |
| 1.5 | 0.015 | 0.985222 | 3.42E+01 |
| 2 | 0.02 | 0.980392 | 3.81E+01 |
| 2.5 | 0.025 | 0.97561 | 4.35E+01 |
| 3 | 0.03 | 0.970874 | 5.22E+01 |
| 3.5 | 0.035 | 0.966184 | 7.10E+01 |

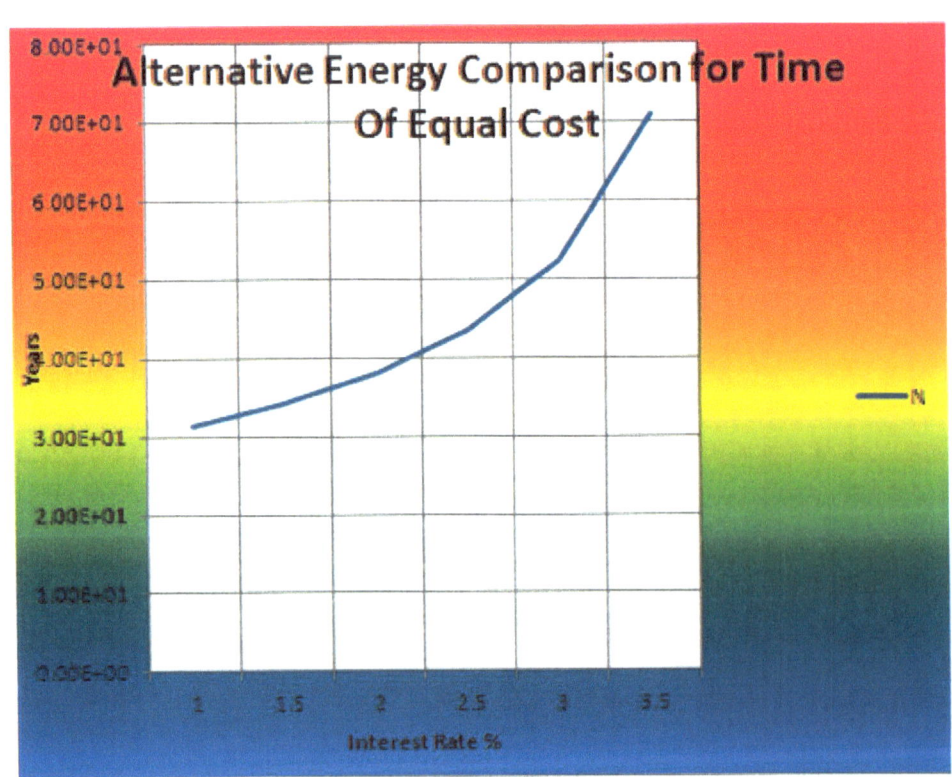

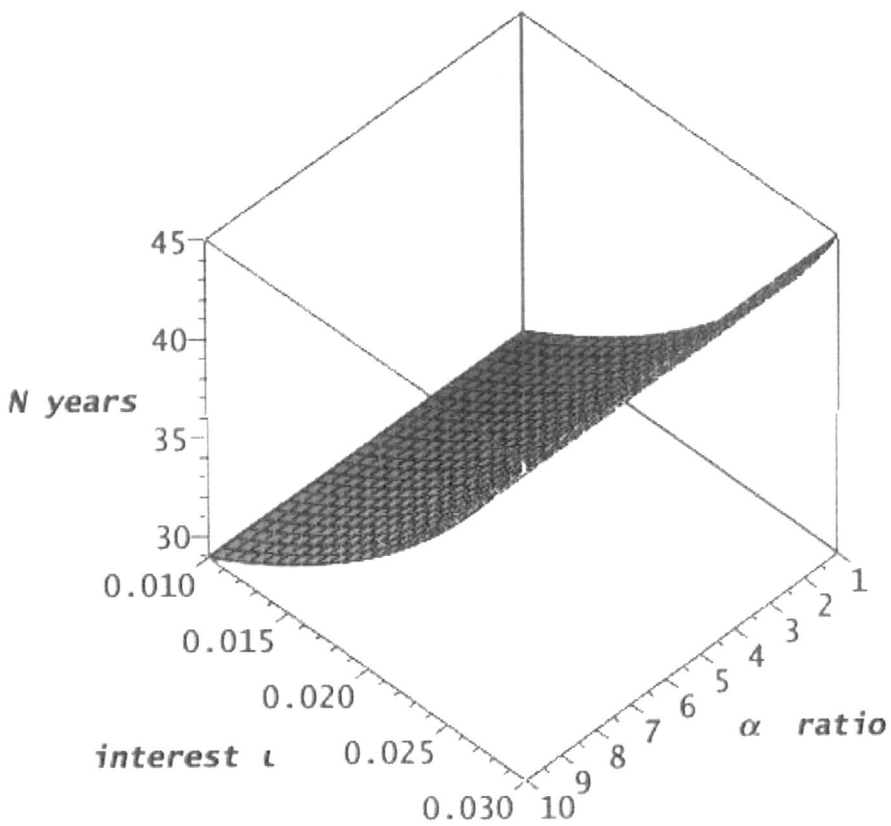

www.ingramcontent.com/pod-product-compliance
Lightning Source LLC
Chambersburg PA
CBHW051023180526
45172CB00002B/451